Aviation and its Management - Global Challenges and Opportunities

Edited by Arif Sikander

Published in London, United Kingdom

IntechOpen

Supporting open minds since 2005

Aviation and its Management - Global Challenges and Opportunities
http://dx.doi.org/10.5772/intechopen.77103
Edited by Arif Sikander

Contributors
Anderson Correia, Marcelo Ramos Martins, Michelle Carvalho Galvão Silva Pinto Bandeira, Eny Yuliawati, Ursula Silling, Tunde Fabgbemi, Adebukola Daramola, Arif Sikander

Notice
Statements and opinions expressed in the chapters are these of the individual contributors and not necessarily those of the editors or publisher. No responsibility is accepted for the accuracy of information contained in the published chapters. The publisher assumes no responsibility for any damage or injury to persons or property arising out of the use of any materials, instructions, methods or ideas contained in the book.

First published in London, United Kingdom, 2019 by IntechOpen
IntechOpen is the global imprint of INTECHOPEN LIMITED, registered in England and Wales, registration number: 11086078, The Shard, 25th floor, 32 London Bridge Street
London, SE19SG – United Kingdom
Printed in Croatia

British Library Cataloguing-in-Publication Data
A catalogue record for this book is available from the British Library

Additional hard copies can be obtained from orders@intechopen.com

Aviation and its Management - Global Challenges and Opportunities
Edited by Arif Sikander
p. cm.
Print ISBN 978-1-83880-660-6
Online ISBN 978-1-83880-661-3
eBook (PDF) ISBN 978-1-83880-662-0

We are IntechOpen,
the world's leading publisher of
Open Access books
Built by scientists, for scientists

4,100+
Open access books available

116,000+
International authors and editors

125M+
Downloads

Our authors are among the

151
Countries delivered to

Top 1%
most cited scientists

12.2%
Contributors from top 500 universities

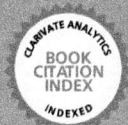

Interested in publishing with us?
Contact book.department@intechopen.com

Numbers displayed above are based on latest data collected.
For more information visit www.intechopen.com

Meet the editor

Dr. Arif Sikander is currently with Murdoch University, Australia. His diverse aviation qualifications include Bachelors, Masters, and PhD degrees in Aeronautical Engineering, Aircraft Maintenance, Flying Studies, Aviation and Airport Management, and Education. His aviation experience spreads over air force (on military jet aircraft), general aviation (on light aircraft), and commercial airlines (wide-bodied aircraft). He holds "Fellowships" of the Institute of Engineers, Australia, and the Australian Institute of Management. Dr. Sikander has published in many reputed journals and delivered keynote addresses. He is a member of the editorial boards of several international journals and has taught aviation courses at vocational, college, and university levels. He has provided consultancy on many aviation projects and is concurrently supervising Master and PhD students. He shared his aviation expertise with the world while being interviewed by Bloomberg on the missing Malaysian flight MH370.

Contents

Preface

The inevitable growth in aviation is attributed to a variety of factors. The rapid growth of countries, especially China and in the Asia Pacific region, provides great opportunities in this sector. Globalization has also contributed to the growth in air traffic, both passenger and cargo. Aviation growth is not restricted to the air transportation sector alone; rather it spreads equally in all its constituent areas, including flying, maintenance, training, marketing, etc. UAV and drones present yet another new sector in aviation, which has great growth potentials. There is a dearth of texts on aviation that address all the areas within this sector. This book attempts to address some of the major issues, provides an in-depth view in terms of the challenges and opportunities, and includes issues such as airline management and operations, airline business models, airport systems, flight operational procedures, aircraft maintenance, runway safety management systems, and air traffic management. Subsequent series of this text will address the remaining issues.

This book contains original research backed by case studies and analytics. The chapters were invited from authors from different countries to provide a global perspective. Chapter 1 introduces aviation as it is at present and the growth potential for the future. Chapter 2 is an analysis of air traffic in a domestic network and shows the pattern of distribution over the last decade. The factors responsible for the observed pattern have been critically explored to determine the demand for air travel. Chapter 3 attempts to examine parameters such as hub airports for international flights in ASEAN (Association of Southeast Asian Nations) that have network efficiency performance. This analysis will be useful to the aviation community, especially with the implementation of an open-sky policy and ever-increasing passenger growth. Runway overrun, which occurs when aircraft exceed the limits at the end of the runway, is an event of interest in the current study. Chapter 4 aims to present an accident model with a new approach in aeronautical systems, based on the tasks of the pilots related to the operational procedures necessary for approach and landing to obtain the chain of events that lead to this type of accident. The latest technology developments such as artificial intelligence, machine learning, blockchain, voice, and more create opportunities never seen before. However, the aviation industry to a large extent has remained stuck in legacy processes and its decades old technology. Finally, Chapter 5 explains the reasons behind the key pain points of the industry, what activities are ongoing, and the main areas that need to change to get into shape for the current dynamic environment.

This book will be extremely useful for students undergoing aviation courses at university level and for airline pilots/engineers, airline/airport managers, and planners. It should also help to open doors and discussions on future research in this area.

Dr. Arif Sikander
Murdoch University,
Australia

Chapter 1

Introductory Chapter: Managing an Unprecedented and Extraordinary Growth

Arif Sikander

1. The stage

Air travel remains a large and growing industry. There is no doubt that aviation has helped accelerate economic growth. With a booming population in Asia and with more reliable and comfortable aircrafts in the air, air travel is expected to rise exponentially across the world and particularly Asia. Flying which used to be a myth is now becoming affordable and as such growth in aviation is extraordinary [1].

The entry of China in the development of new aircraft, e.g. C919, will bring the competitive field to a playing level which is currently dominated by Airbus and Boeing. This competition is supposed to reduce the cost of travel further and make air travel more affordable, thus further accelerating the growth. The growth in the number of flights and as such the aircrafts will also result in the growth of airports and jobs. This will bring a radical change in the aviation industry. The government and the private sector will see new opportunities and challenges. This will require new sets of management tools and expertise and provides opportunities in academics to develop such instruments to exploit the opportunities.

2. Contributory factors

The inevitable growth in aviation is attributed to a variety of factors. A major factor contributing to this has been the rise of the middle class in countries like Brazil, Russia, India and China (BRIC). The affordability of the middle class has supported the airline industry in terms of leisure and business travel. This has been mainly due to reduction in the cost of air travel among intense competitions. The low-cost airlines has been a vital source in making this travel possible for those who never dreamt of travel.

Improvement in safety due to more reliable aircrafts, engines, aircraft components has eliminated accidents and provided impetus to air travel by people in villages and rural areas who were fearful to travel in the past. The fatality rate has been drastically reduced making scheduled passenger airline service more safe. The year of 2017 was safest year for air travel as fatalities fell to 79 deaths compared to 1000 deaths in 2005 [2].

Increase in routes due to both reliable and fuel-efficient short-distance and long-distance aircraft availability has provided impetus to this growth in air travel [3]. The establishment of industries by the USA and other countries in China has also been a major factor in terms of establishing new routes to and from China for the people from these countries working over there.

With the financial crisis still looming over many countries, tourism has been adopted as a tool for economic growth. This has tempted people to travel to countries (affordable travel cost) which they would have avoided previously. Even the countries which either have not seen the financial crisis or are well-off have been marketing tourism, thanks to the availability of new routes and the cheaper airfares. Many long-distance direct flights are being planned in the coming years towards tourism to boost their economies.

Another factor providing boost to this sector is the code sharing. Major airlines are partnering with domestic airlines to fly their customers to local destinations.

3. The future

By 2025, annual air travel is predicted to double with Asia-Pacific region dominating and China is expected to leave the USA behind in less than 10 years [1]. Such a tremendous growth will imminently result in the increase of air traffic. With air traffic growth, issues like airspace congestion, airport congestion and greenhouse emission need to be carefully looked into by all stakeholders. There will be a pressure on the governments to achieve emission control targets set by Carbon Offsetting Reduction Scheme for International Aviation (CORSIA). This growth in air travel and airports will place more demand on real-time security. With the recent incidence of Gatwick airport being inoperative with hundreds of flights cancelled due to drone sightings, a new challenge on "obstacle data management" has emerged [4]. A surge in demand for aviation personnel, pilots, engineers, cabin crew, etc. would need investments in the training systems. Research in developing better and greener fuels for the jet engine will need to be enhanced. Unfortunately due to a shortage of the human resources in this sector, it is envisaged that a lot of efforts would need to be expended in the training and development of these human resources. This sub-sector (aviation training) provides a huge challenge and opportunity.

4. Conclusion

Aviation growth provides both social and economic benefits. The demand of air services increases the influence of air transport in the global economy, thereby providing opportunities for people and goods movements ([5], p. 4). The growth of aviation has spillover effects as well, thereby uplifting other sectors of the economies. Aviation brings economies and people nearer and as such brings communities together, thereby enhancing cultural cohesion. Air travel is by far the most preferred mode and is no more a dream. The predicted growth of aviation sector over the next decade has provided both opportunities and challenges for the governments, airlines, and other stakeholders. It is imperative that academics explore and develop new tools in 'business and management' specifically for this sector so that this complex activity is understood and managed effectively. The list of issues to be explored is huge, but the important ones include airline marketing, airline management, airline economics, airport and airlines, environmental and sustainability issues, green and smart airports, climate change and aircraft performance, technology and automation, cargo management, air traffic management, air capacity constraints, airport congestion, new airport designs, aviation safety, new large aircraft design, small aircraft transportation systems, drones and their impact, advanced flying techniques, real-time condition monitoring, global aviation data management, passenger comfort enhancing new aircraft systems and new integrated training systems. This book is an attempt to visit these issues and is the first one in the series to follow.

Author details

Arif Sikander
Murdoch University, Australia

*Address all correspondence to: arif.sikander@murdoch.edu.au

IntechOpen

References

[1] The Hindu Business Line. India's aviation growth is unprecedented: Jayant Sinha. Mar 15, 2018. Available from: https://www.thehindubusinessline.com/economy/indias-aviation-growth-is-unprecedented-jayant-sinha/article23263853.ece

[2] BBS News. 2017 safest year for air travel. Aired on 2 Jan 2018. Available from: https://www.bbc.com/news/business-42538053

[3] Rosen E. As billions more fly. Here's how aviation could evolve. National Geographic. 20 June 2017. Available from: https://www.nationalgeographic.com/environment/urban-expeditions/transportation/air-travel-fuel-emissions-environment/

[4] Petrovsky A, Doole M, Ellerbroek J, Hoekstra JM, Tomasello F. Challenges with obstacle data for manned and unmanned aviation. The International Archives of the Photogrammetry, Remote Sensing and Spatial Information Sciences. Volume XLII-4/W10; 2018

[5] ATAG. The social and economic benefits of air transport. Air Transport Action Group; 2005. Available from: https://www.icao.int/Meetings/wrdss2011/Documents/JointWorkshop2005/ATAG_SocialBenefitsAirTransport.pdf

Chapter 2

Air Travel and Airline Operations in Nigeria: Market Potentials and Challenges

Adebukola Daramola and Tunde Fagbemi

Abstract

In this chapter, we analyse air traffic in Nigeria's domestic network and show the pattern of distribution in the last decade (2008–2017). We attempt to analyse factors responsible for observed patterns, thus explaining contributory factors driving the demand for air travel. Air fares are a major factor in demand, as there are alternative modes by which journeys can be completed. High costs related to provision of airline services appear to be a major factor driving prohibitive air fares. Much of the potential market for air travel is thus excluded to other modes, while traffic patterns show few commercially viable nodes—Lagos, Abuja and Port-Harcourt in a network of 20 cities. The paper proposed strategies to achieve reduced cost of operations for airlines, which hopefully will inform reduced fares and positively affect propensity to fly in Nigeria.

Keywords: network, air traffic, Nigeria, passenger, propensity

1. Introduction

The air transport industry in Nigeria experienced a landmark event in 1985 when entry into the domestic airline sector and air fares were formally deregulated. The narratives around the industry gradually changed from a national carrier focus to one focused on private carriers. By 2003, the industry was composed entirely of private carriers. Apart from deregulation of airline services, government, in recent years is gradually divesting from air transport infrastructure and encouraging provision of these by the private sector as well. Some gains of private participation are readily observable in the airline sector, namely increased competition, more efficient electronic based booking systems, and discount fares in some cases. Analysis of air transport in Nigeria easily bifurcates into pre-deregulation and post-deregulation periods; however, after more than three decades of deregulation, patterns of air traffic in the post-deregulation period can be interrogated for meaningful explanations without much recourse to the pre-deregulation years.

The air transport subsector in Nigeria accounts for the second highest share of modal contribution to transport output. The road sub-sector accounts for as much as 84% of transport GDP while air transport share in the last couple of years has averaged about 6–7%. With respect to contribution to the local economy, Phillips Consulting [1] notes that the aviation industry supports 254,500 jobs in Nigeria and contributes US$940 million (N184.7 billion) to national GDP13. Of this sum,

49% (i.e. US$462 million or N90.8 billion) is a direct output of the aviation sector (via airports, airlines and ground services), while the remainder is acquired indirectly (via the supply chain). An additional US$464 million (N91.2 billion) is derived from tourism, which raises the overall contribution to US$1.4 billion (N275.9 billion). Nevertheless, the air transport sub sector has much more potential for contributing to the local economy particularly through increased capacity for earning and conservation of foreign exchange.

The industry has witnessed a high turnover of domestic carriers since deregulation; generally, many of the local airlines experiences in the industry have been short lived, with many often operating a few years and then folding up. Currently there are only about eight active scheduled domestic passenger carriers in the Nigerian airline industry, although total number of active Nigeria registered carriers is put at 23 [2]. In the next section we discuss air travel demand elasticities as a guiding concept for the discourse.

2. Air travel demand elasticities

The demand for a particular good or service depends on a variety of factors, among them consumers' taste, income level, price and quality of the product in question and the prices of other goods, especially goods that are close substitutes. As a general rule, when other influences on demand remain unchanged, a higher price for a product results in a lower quantity demanded. However, the price responsiveness of demand varies from one good to another and from one market to another. The own-price elasticity of demand measures the responsiveness, or sensitivity, of the demand for a good to changes in its price when other influences on demand are held constant. It is defined as the percentage change in quantity demanded resulting from a given percentage change in price.

In the case of air travel, studies distinguish among markets for business and leisure travel; as well as for long-haul and short-haul travel. Accordingly, to examine the sensitivity of the demand for air travel to its price, separate estimates of the own-price elasticity of demand are gathered for each of these distinct markets [3].

Since the availability of alternative modes of transportation that are reasonably close substitutes for air transport diminishes with distance travelled, it is expected that the demand for air transport will be less elastic for longer flights than for shorter flights. Furthermore, international travel tends to be spread over more time than domestic travel, so that the airfare is a smaller proportion of overall trip costs, which makes international travel less sensitive to changes in ticket prices. In addition, leisure travellers are more likely to postpone trips to specific locations in response to higher fares, or to shop around for those locations offering more affordable fares. Consequently, it is expected that the demand for air transport for leisure reasons will be more elastic than business travel.

This basic concept of own-price elasticity of air travel in different market segments suggests that if air fares are reduced on Nigeria's domestic routes, demand for air travel is likely to increase, since these routes are short-haul.

The next section examines revenue passenger kilometres achieved vis-a-vis economic and demographic profiles for Nigeria and selected comparator countries.

3. Cross-country profiles

Comparator countries are chosen based on similar geographical size range (Venezuela and Egypt), similar demography, specifically population size range (Brazil and Pakistan) and similar economic profiles specifically size of GDP (Venezuela and Egypt).

Nigeria achieved the lowest revenue passenger km (1894 km) in the 2015/2016 period among the five countries. Venezuela's population is only about 17% of Nigeria's but, it achieved higher revenue passenger km (5142 km) in 2015/2016 period. Similarly, Egypt, which is only about half the Nigerian population, achieved revenue passenger km eight times higher than that of Nigeria during the period. Pakistan, whose population is closely similar to Nigeria's recorded revenue passenger km seven times higher than Nigeria's. Although real GDP growth rate for Nigeria is highest among the five countries, its GDP per capita is lower than what obtains in other countries, except Pakistan. Aside from Brazil which is much larger, other comparator countries are within similar territorial size range as Nigeria.

As seen in **Table 1**, Nigeria has the lowest propensity to fly among all countries for which data was available. This story is a paradox of sorts, given that the geography as well as the demographic profile in Nigeria favours air travel. The country has a working population of over 73 million, which, in addition to the fact there are substantial inter-city distances, should favour propensity to travel by air. The low GDP per capita probably provides some explanation for low PTF, but again, Pakistan has a lower GDP per capita and still manages to record a higher PTF than Nigeria. The number of active domestic airlines is also lower in Nigeria than in other countries, again indicating the low level of demand for air travel.

Air fares are observed to be on the high side. The most trafficked route in the network, Lagos-Abuja has an average fare of N30,000 per passenger flight hour. This translates to about $83.3 at the current rate; meanwhile, flights on Boeing 737–700 series in western countries offer $33.33 per passenger flight hour [8]. In Africa generally, infrastructure services are observed to be twice as expensive as elsewhere. This is not peculiar to air transport; power, water, road freight, mobile telephones and internet services also mirror the same trend.

Customer confidence in Nigerian airlines is another reason air travel demand is deemed low. The aircraft stock shows that the average fleet age is about 20 years. This contrasts sharply with fleet age for Africa's best airlines—Ethiopia airlines for example is said to have an average fleet age of 5 years.

The government of Pakistan in 2015 launched a 'liberal bilateral open skies ational aviation policy'. This policy included incentives to investors, among them

Country	Air traffic indicators (2015/2016)			Geography/demography (2016)		Economic Indicators (2008–2016 average)		
	Rev pax (million)	PTF[1]	Number of active domestic airlines	Territory ('000 km²)	Population (million)	GDP (USD billion)	GDP per capita	Real GDP growth (%)
Nigeria	3373	0.018	8	923	183.6	318.7	1894	4.7
Venezuela	5142	Na	21	916	31	305.8	10,326	1.1
Brazil	120,001	0.5	16	8514	206	2032.5	10,268	1.6
Egypt	23,180	0.114	16	1001	90.2	253.5	3684	4.3
Pakistan	21,311	0.043	Na	881	181.7	221.5	1241	3.5

Sources: [4–7].
[1]PTF refers to propensity to fly. It is calculated as volume of passengers divided by country population in the year.

Table 1.
Country profiles.

a zero taxation on investments in the sector. A minimum of eight aircraft was set as threshold for any airline company willing to operate in the country- three for domestic and five for international operations. Service provider functions were also separated from regulatory functions [8].

Venezuela runs a socialist economy, hence the aviation industry is mostly government controlled. Hyper inflation, macroeconomic distortions coupled with state intervention have contributed to volatile regulatory framework in the country. Nevertheless, Venezuelan domestic carriers have struggled to keep afloat, swapping domestic flights for international services [9]. Despite these fundamental challenges, Venezuela achieved higher revenue passenger km than Nigeria in 2015/2016 period.

In the following section, we examine spatial patterns and nodal traffic densities in Nigeria's air transport network.

4. Airline network patterns

According to the NCAA, Nigeria has 20 airports, over 18 aerodromes and over 30 regulated airstrips and heliports; 23[1] domestic airlines; 554 licenced pilots; 913 licenced engineers and 1700 cabin personnel. Nigeria is an important destination for over 22 foreign carriers, it currently has Bilateral Air Services Agreements with over 78 countries. There are direct connections from Nigeria to many of the world's business centres such as London, Paris, Frankfurt, New York, Johannesburg, Atlanta, Amsterdam, Dubai and Jeddah to mention a few. With the attainment of America's Federal Aviation Administration (FAA) International Aviation Safety Assessment (IASA) Category One Certification in August 2010, Nigerian registered carriers can now fly directly into the United States of America (USA) [2]. Above provides a background for observed network patterns. Our emphasis in this chapter is on the domestic network.

The domestic network, which is naturally built around the airports has changed configuration over the years. In 2008 there were fewer air corridors for traffic as seen in **Figure 1**.

By 2017, the network had a higher route density, showing that new air corridors were cultivated over the period (**Figure 2**).

During the 2008–2017 period, 139 million passengers flew through Nigeria's airports, 100 million of these were domestic passengers, while the rest were International passengers. Domestic passengers formed over 70% of passenger traffic during the period. Growth in domestic passenger traffic ranged from −14% in 2017, to as high as 39% in 2008. Growth trend was mixed during the decade, recording negatives in 2012, 2013, 2015 and 2017. There was a major crash in the domestic sector in 2012 which dampened demand in the ensuing months. 2015–2017 were years of economic slow down. The first signs of recovery occurred in late 2017 from an economic recession, which reached a bottom out point in 2016. These among other factors were responsible for declines in passenger traffic in various years.

During the entire decade, Lagos, Abuja and Port-Harcourt were the nodes with the greatest shares of traffic. Kano, Kaduna, Enugu and Osubi had medium volumes of passenger traffic, similar to Benin and Owerri in 2008/2017. Other nodes in the air network were small volume nodes. This leads us to discussions on nodal densities in the network.

---

[1] Not all of these are active, some do not offer scheduled passenger services.

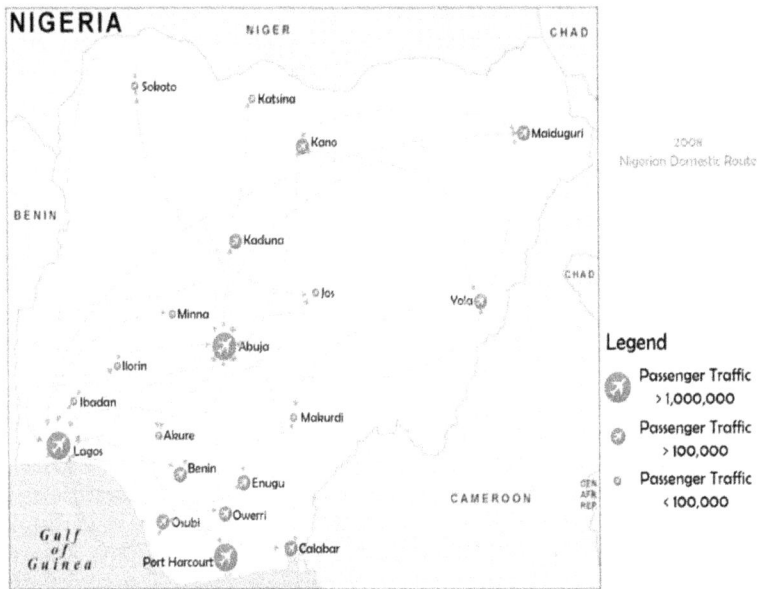

Figure 1.
Air traffic network, 2008. Source: Authors' survey.

Figure 2.
Air traffic network, 2017. Source: Authors' survey.

4.1 Nodal traffic densities

Airlines in Nigeria, as in several places, are mostly passenger movers, hence, our focus is on passenger traffic. Traffic at the 20 nodes shows that over the period 2007–2016[2] Lagos, Abuja and Port-Harcourt airports accounted for 76% of domestic

---

[2] Traffic was disaggregated by nodes for this period.

passenger traffic at the 20 domestic airports. Shares of passenger traffic in the three cities were 36, 30 and 10% respectively. This pattern shows there is strong dominance reflected in a sharp core-periphery structure of air passenger movements. Abuja and Lagos are Nigeria's political and commercial capitals respectively, while Port-Harcourt is a major oil producing city. Clearly, the cost if air fares naturally excludes a large share of Nigeria's travelling public.

Much of the movements recorded in Lagos pertain to corporate travellers in the middle and high income categories, Lagos houses much of this group in Nigeria given its status as a megacity. In Abuja, passengers are mostly top government and private sector workers, while Port-Harcourt travellers thrive on the oil economy. The lower middle class where a great potential for market exists generally do not find air fares affordable. They therefore resort to corporate road transport services.

The implication of low traffic densities in several nodes is that many city-pair routes are not commercially viable to the degree that active airlines will increase their service frequencies on these sectors. Consequently, many nodes in the network do not record sufficiently large passenger movements. Nevertheless, city pairs in Nigeria's network have great potential for air travel as road distances on these corridors range from 200 km to over 1400 km. Air transport offers the fastest means of covering these distances as long as airlines keep to scheduled departure time.

5. Factors constraining airline operations and viability

5.1 Operational/maintenance costs

Airlines maintenance costs are mostly borne in foreign currency. Scarcity of forex for airlines operating in the country was even more pronounced in the wake of the recession in the 2016/2017 period. The exchange rate of the Naira to the dollar increased by over 200% at a point and this increased airlines' operational costs significantly. Taxes are equally high, so is fuel cost.

5.2 Small size of carriers

Nigerian airlines are small, with fleet sizes as low as three for some airlines, the actual market is equally small. Although market potentials exist along several under-utilised air corridors, the smallness of airlines does not permit them to explore these potential routes. Airlines may not be able to break even given the low load factors that are likely to exist on such routes. Small size of carriers also constrains capacity to offer frequencies and compete on regional and international routes. Nigeria's domestic airlines are therefore not strong players in the international and regional markets.

5.3 Lack of airline competition

Competition is a natural result of many players in an industry. Economies of large scale production naturally drives down costs. However, as profit oriented actors, airlines will not venture into an industry where there is low propensity to fly. The lack of airline competition and the absence of regional airport hubs are some of the constraints identified in Africa's aviation [10].

5.4 Infrastructure challenges

Power is one of Nigeria's biggest infrastructure challenges. The cost of providing alternative sources (usually using diesel powered generators) in the face of public power shortages is prohibitive. The airport operators naturally pass these costs on as part of airport charges. Airlines will then pass these on to the final consumers. These all add up to airfares.

6. Recommended strategies for improved airlines' operations

6.1 Reduce airlines' maintenance costs

A worthwhile strategy would be to put policies in place which will enable airlines reduce cost of operations. Thereby, lower fares can be offered to attract more patrons into air travel. Establishing an aircraft leasing hub and aircraft maintenance, repair and overhaul facility in Nigeria will be a step in the right direction. If domestic airlines can access aircraft and maintenance services locally, then scarce foreign exchange can be conserved. New airlines venturing into the industry may also be given incentives in the form of some tax holiday for specified period. The responsibility here lies with the regulatory agencies—Nigerian Civil Aviation Authority (NCAA) and Federal Ministry of Aviation (FMA).

6.2 Explore airline co-operation

There may be economic sense in exploring co-operations at the national as well at the regional level. Such arrangements enable players to create and take advantage of scale economies. Hopefully, competitive prices and better quality will result and will trigger increased demand for air travel. Amalgamations will also enable domestic carriers compete better in the regional and international markets. Co-operations and alliances have proved useful among air carriers in developed climes. Incidentally, carriers in developed countries are several times bigger than African carriers. The responsibility here lies with airline operators and management.

6.3 Provide alternative power infrastructure

It may be worthwhile to consider renewable energy options, such as solar power to complement power needs at the country's airports. Nigeria falls within the tropics and has average daily sunshine time up to 8 h in dry season months. Although the capex for solar power facility may be high, subsequent recurrent costs are likely to be far below costs incurred from current power provision options being explored. This could be another way of achieving reduced costs of airport operations and of airport charges passed on to airlines, which eventually gets passed to consumers. FMA and Federal Airports Authority of Nigeria (FAAN) will bear responsibility for this action.

7. Conclusions

Clearly, increased demand for air travel and improved air transport operations need to be engineered in Nigeria. While the demographics and geography are highly favourable for air travel, the prohibitive costs of air travel exclude several potential consumers of the service. The high cost of air travel is traced to supply side costs of operations, maintenance, taxes and other regulatory charges. These supply side costs can be reduced if the strategies proffered are explored. Recommendations include tax holidays to incentivize new entrants, co-operations among airlines, establishment of a local AMRO to conserve foreign exchange spending by airlines and provision of alternative renewable power to cut down cost of airport operations. Hopefully, these supply cost reduction measures will result in reduced/more competitive fares and incentivize potential consumers to patronise air travel. We hope by this, Nigeria's propensity to fly will increase, higher revenue passenger kilometres will be achieved and air transport can contribute more significantly to transport output in the nation's economy.

Acknowledgements

The authors acknowledge the contributions of Ramat Balogun, Emeka Ogri and Busayo Tejuosho who helped with data gathering and literature review.

Conflict of interest

The authors affirm that this research paper has not been previously published in other outlets.

Author details

Adebukola Daramola[1*] and Tunde Fagbemi[2]

1 Nigerian Institute of Social and Economic Research, Ibadan, Nigeria

2 Spring Fountain Infrastructure Limited, Lagos, Nigeria

*Address all correspondence to: adebukoladaramola@gmail.com

IntechOpen

References

[1] Phillips Consulting: The 2015 Domestic Aviation Industry. Customer satisfaction survey report. 2015. www.phillipsconsulting.net

[2] Nigerian Civil Aviation Authority (NCAA). NCAA and ICAO sub-section on NCAA website (homepage). 2018. www.ncaa.gov.ng

[3] Department of Finance, Canada. Air travel demand elasticities: Concepts, issues and measurements 1. 2008. www.fin.gc.ca

[4] Klaus Schwab. The Global Competitiveness Report, 2009-2017. World Economic Forum. Geneva

[5] Central Intelligence Agency. The World Factbook. 2016. Central Intelligence Agency. www.cia.gov

[6] The World Bank. World Development Indicators. World Bank, Washington DC; 2016

[7] International Civil Aviation Organization. Presentation of 2016 Air Transport Statistical Results by ICAO, 2016

[8] Roberto Leiro. Best of Airways-Venezuela's Turbulent Skies. 2017. https://www.google.com/amp/s/tribune.com.pk/story/856566/forward-looking-policy-skies-make-way-for-new-aviation-policy/%3famp=1

[9] Available from: https://airwaysmag.com/industry/whats-next-aviation-venezuela/

[10] Foster V, Briceno-Garmendia C. editors. Africa's Infrastructure: A time for transformation. Agence Francaise de Developpement and The International Bank for Reconstruction and Development/The World Bank. Washington, DC, USA; 2010

Determination of the International Hub Airport to Support the Flight Network Efficiency of ASEAN Region Countries (Case Study of the Indonesian Airport System)

Eny Yuliawati

Abstract

Indonesia has 266 airports spread throughout the Indonesia archipelago. With the growth rate of air passengers increasing year by year, Indonesia needs to increase its role in managing the international network in Southeast Asia. Especially with the implementation of the Open Sky policy, Indonesia must take advantage of the potential opportunities. This chapter attempts to examine parameters at hub airports for international flights with the Association of Southeast Asian Nations (ASEAN), which has network efficiency performance. There are eight airports actively showing behaviour as "hubs", and considering the potential geographic and movement potential in ASEAN, the most efficient is the three hubs and 32 spokes configuration. Thus, the three hub airports that can be optimised to support the efficiency of international flight routes in ASEAN are Kualanamu Airport-Medan, Soekarno Hatta Airport-Jakarta and Juanda Airport-Surabaya.

Keywords: hub and spoke airport, flight network efficiency, international flight route, ASEAN region, open sky policy

1. Introduction

The Association of Southeast Asian Nations (ASEAN) Open Sky is an important component in ASEAN economic integration. It is based on the idea that the transport network, especially air transport, is very important in facilitating changes and reducing trade barriers. The important role of the air transportation sector becomes a decisive factor considering that the field of tourism is a major stimulus for economic growth for some countries in ASEAN. The Open Sky policy is a waiver of the rules for the international aviation market to minimise government intervention on provisions applicable to scheduled airline and charter markets, both passenger and cargo flights, where their implementation is based on bilateral and multilateral agreements. An important element in Open Sky is free-market competition where there are no longer any restrictions on the rights of international flight routes (the number of flights, capacity, frequency and type of aircraft), while prices are determined by market power as well as providing equal and fair opportunity to all airlines to compete. For that reason, Open Sky in the ASEAN region will encourage the airline industry to become more competitive in providing places for all flights from ASEAN to compete in intra-ASEAN region routes as well as providing additional flight flexibility for route (network) expansion. Furthermore, the presence of Open Sky in the ASEAN region is expected to encourage airlines to be more efficient, which in turn allows airlines to reduce unnecessary aviation operational costs so that low-cost carrier flight models will become better developed in competitive conditions.

There are two levels of measurement that can be used to determine flight efficiency, namely network and infrastructure capacity measurement. Network measurement can be done on the basis of a network structure. At this level, minimisation of the "cost" that is caused is related to the variables of travel time, distance, connectivity, etc. Meanwhile at the infrastructure capacity level, measurements are made based on airport infrastructure capabilities in serving demand such as runway, apron, terminal and other capacities, for example, in minimising the "cost" based on "connectivity" in calculating connectivity on transportation networks using algorithmic engineering so that it can find the fastest route between two points "s" and "t" [1]. Reducing the costs of travel and increasing their connectivity are major advantages of the hub and spoke network system [2]. Connectivity is increased within the hub by concentrating landings and takeoffs at the hub (hub waves) [3]. Furthermore, related to the network pattern in flight, there are "hub" and "spoke" flight network patterns where all flights head to one large central location and passengers can transfer to other flights to reach their final destination. The pattern of hub and spoke development has been used by commuter groups in the United States since the early 1980s. This pattern has been able to expand and organise the route network and could prioritise the interest of consumers or air passengers. The arrival of airplanes in the "hub" and "spoke" flight network pattern is well coordinated, and to make it easier for passengers or goods that are to be transferred to another flight, this pattern is repeated several times a day [4]. The hub airports serve as a consolidation of passengers and cargo that can move to other airports categorised as spoke airports and provide a connecting flight to various subsequent destinations. The selection of hub airports is based on strategic geographic factors and demand. For this reason a method of planning and route optimisation is used so that the system can be planned accurately.

Airports are categorised as hubs and spokes based on freight ratio, which is the ratio between the weight of goods and the number of boarding passengers served at the airport. In addition to establishing airports as hubs it is also based on consideration of passenger traffic density, airport geographic location, airport area, supporting transportation mode facilities, short and long distance flight traffic flows, the number of "banks" (grouping on a daily frequency of arrival/departure in several terms) and "bank" operating period [5]. The number of banks and the period of bank operations are determined by passenger air traffic, including demand derived from spoke passenger routes, arrival timings and departure of long-haul destinations and the choice of passenger based on flight scheduling.

Furthermore, with the implementation of the ASEAN Open Sky policy, there are a number of problems, such as:

1. There may be a change in determining the location of the airport as a hub airport (transshipment airport).

2. What is the ASEAN Open Sky can be promote the aviation industry competition and supporting the airline industry to doing competition inter ASEAN region better.

Although the largest increase in air transport movements in Indonesia is domestic flights, the implementation of the Open Sky policy will encourage international flights to grow significantly, therefore it is necessary to study the establishment of international aviation hub airports (transshipment) to improve the efficiency of flight routes to the ASEAN region.

2. Open sky policy in ASEAN

The definition of Open Sky refers to the situation of broad liberalisation in the ASEAN region and concentrates on international relations of ASEAN member countries, which will provide additional flight flexibility for route development. ASEAN Open Sky is a set target in the "Road Map for ASEAN Integration: Competitive Air Service Policy" prepared by the ASEAN Air Transport Working Group [6].

Each country has different policies. There are some countries that are very liberal but place limitations on international flights. There are also some countries that choose to gradually take steps towards liberalisation, and there are also countries that are ready to support liberalisation on a subregional basis. ASEAN members have diverse characteristics with respect to air transportation based on the levels of growth and development [7–11]. Regarding the implementation of the Open Sky policy, there are different types of agreement, such as bilateral agreement or multilateral agreement, that are used for supporting various policies in the ASEAN Region. The following are three subregions in the Open Sky policy in the ASEAN region [12]:

1. The first subregion is Cambodia-Laos-Myanmar-Vietnam plus Thailand and Brunei Darussalam

2. The second subregion is Vietnam-Indonesia-Philippines plus Brunei Darussalam; these countries have the same progress in aviation industry development.

3. The third subregion is Singapore-Malaysia-Thailand plus Brunei Darussalam; this subregion is used in countries that already have regular flights such as Singapore, Malaysia and Thailand with the possibility of Brunei Darussalam joining them.

3. Network planning and flight path

The counting of connections on the transport network performed by the application of algorithmic engineering. A problem can be solved by using transportation network modelling in a graph that describes the travel time on the trip in question. The algorithm can solve the problem by finding the fastest route between two points "s" and "t". The challenge is to determine the appropriate transport model to use in the graph. While road networks can be modelled simply (intersections as nodes, roads as links), realistic modelling on public transportation networks is far more complicated. [1].

Figure 1 describes the time-dependent model that determines a flight path; the flight of "F" is divided by a number of "R" routes. In this case, two F1, F2 and F flights are considered equivalent if they share the same airport sequence (airport (A1, ..., Ak)). The curve is built using every airport point and route through a particular airport, following the illustration of a simple example of the time-dependent model. The flight schedule for all routes has a length of 1, because almost all flights do not make a temporary stop. The flight schedule is assumed to be a direct flight for each pair of airports (Ai, Aj) with i < j; the result is that all routes have a length of 1.

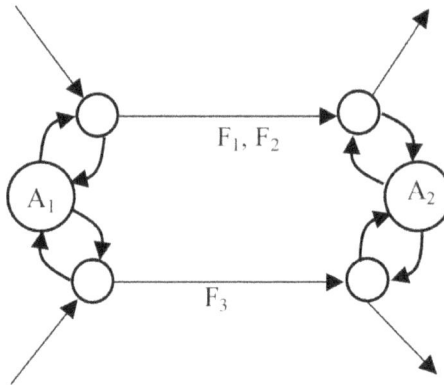

Figure 1.
Time-dependent model of the determination of a flight path.

4. Airport planning for hub and spoke pattern

The hub and spoke flight network pattern arises when all flights are heading to a large central location, passengers change flights to reach their final destination and this pattern is repeated several times a day. This pattern consists of several hub airports that function as centres of economic activity and flight activities in a region surrounded by small towns that will be in direct contact with them (**Figure 2**).

The hub and spoke pattern is not new to the aviation world. It was first introduced and developed by commuter groups in the United States in the early 1980s. This hub and spoke system has been able to develop and organise routes, as well as promote public and consumer interest which done by the trunks community (the later known as the US Majors) and then followed by the locals group (the later known as US Nationals). Progress was triggered by implementation of Airline Deregulation Acts in 1978.

Thus, in this model the flight route consists of a central point (hub) that serves multiple ends (spokes). The hub serves as a consolidation of passengers and cargos that are transferred to the various spokes and provide the connecting flights for the next destination, both domestic flights and international flights. Airline operators organise interhub flights several times a day, usually using large capacity aircraft that can carry passengers from areas (districts) around the hub airport.

Airline operators also organise fleets for the spoke airport using smaller aircraft, providing higher flight frequencies and supporting hub airports by connecting to the large number of spoke airports as well as building partnerships with regional airline operators or establishing branches to build networks to remote areas [13]. **Table 1** shows this pattern based on airlines offering the best connections per arriving flight.

The selection of hub is based on the location and the large market demand for pairing of origin–destination trips by supporting an airline operation. For this

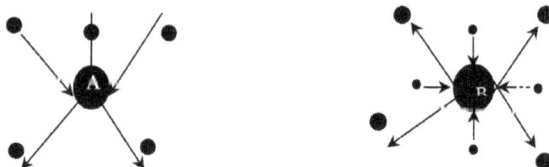

Figure 2.
Hub and spoke for a flight network pattern. A:Hour glass hub & spoke. B: Hinterland hub & spoke.

Airline	Hub	Average total connections per arriving flight	Average rapid connections per arriving flight (%)	Average slow connections per arriving flight (%)
Etihad airways	AUH	19	66	34
Qatar airways	DOH	13	61	39
Air France	CDG	37	59	41
Lufthansa	FRA	55	55	45
KLM	AMS	45	52	48
Emirates	DXB	21	50	51
British airways	LHR	39	47	53
Source: [14].				

Table 1.
Average number of connections per arriving flight at the hub of selected airlines.

purpose, the method of planning and optimising for detailed route determination is used to obtain an accurate basis for transportation system planning.

5. Hub airport identification

To identify the determination of hub airports the Herfindahl–Hirschman Index method can be used. HHI is a distribution and market size measure equal to the sum of squares of participation of a corporation "i" (HHI = $\Sigma \pi^2$). HHI is widely implemented in transportation to analyse the relationship between ticket prices and market concentration and is also recommended by some organisations, including the US Department of Justice and the US Federal Trade Commission [15]. HHI is attractive because it correlates directly with the number of effective market participants. HHI can be considered to be in contrast withthe number of companies that have the same size, so if the HHI in a market has value 0.1, then this market will be as competitive as the 10 companies with the same large size. Symmetrically, the opposite of HHI can be regarded as the number of airports (ne) in the market, which are airports with a significant market share in the system.

The values of HHI should be in the range of "0" to "1"; however, in the case of airports only in the range of 0 to 0.5, every time an aircraft takes off means one is landing at another airport. Therefore, any airport cannot have more than 50% of the movement of aircraft in an aviation network. The maximum concentration of air transport systems with "n" airports occurs when there is only one hub airport with 50% of the market share and the remainder is distributed equally on the spoke airports.

Figure 3 illustrates the dashed blue line at four departure banks, which begins with a low frequency and climbs to a high frequency; the banks begin with a high density until the levels decline, as follows:

1. Bank 1 at 6.30 am to 11.30 am, the peak hours at 8.30 am to 10.00 pm.

2. Bank 2 at 11.30 am to 14.30 pm, the peak hours at 11.30 am to 13.00 pm.

3. Bank 3 at 14.30 am to 19.00 pm, the peak hours at 15.30 pm to 18.00 pm.

4. Bank 4 at 19.00 pm to 23.30 pm, the peak hours at 20.30 pm to 22.00 pm.

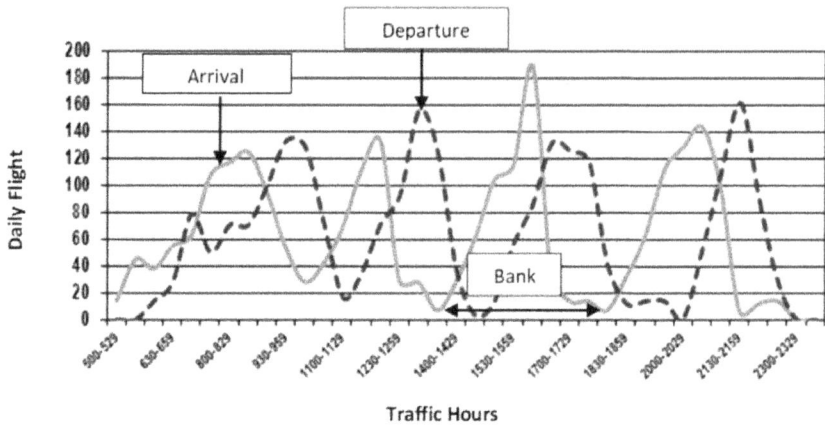

Figure 3.
Grouping of arrival and departure movements for a conventional hub.

The air transport network in Indonesia using the hub and spoke airport system can be assessed using an analysis of the relationship between airports as nodes and attributes, for example the number of routes served at the airport, the number of boarding passengers at the airport and the weight of cargo carried through the airport. If the relationship between nodes and attributes forms a normal distribution curve, then the network pattern system is the hub and spoke pattern system.

There are three main points that are the reference for air transport business corporates and hub airport managers to attract as many passengers as possible:

1. To arrange the wave of arrivals and departures on "banks" from the beginning (morning period) to the end (night period) during the day (during the operational hours), so that the waiting time for passengers was not too long.

2. To set up the flight arrival path of the spoke airport in the long-haul (domestic/international) flight departure path, and vice versa, to make it comfortable for passengers due to GMT time difference.

3. To serve the suburbs and specific markets that cannot be served by a point-to-point system.

6. Analysis of hub airport efficiency for international air traffic in the ASEAN region

6.1 Eigen vector centrality method

Eigenvector centrality (EVC) analysis is performed by assessing the level of importance of an airport in relation to another airport based on airport connectivity in the network system [1]. As a simple illustration, airport A is more important in a network structure if it is connected to airport B (the primary hub airport category) than if it is connected to airport C (the secondary hub airport category). For example, Radin Inten II Airport-Lampung Province will be rated more important if it is provided with the flight route to Soekarno-Hatta International Airport-Jakarta Province (the big hub airport category) compared to the flight route to Sultan Thaha Airport-Jambi Province (the medium hub airport category).

To obtain this level of importance requires the simultaneous assessment of all the airports under observation. Therefore, assessment is done by using a matrix operation, in this case the Origin Destination Matrix, to facilitate an understanding of the form of graphical visualisation as presented in **Figure 4**.

Statistical analysis shows that the distribution pattern of EVC value (Ev) follows the power law pattern when the value of the square coefficient is 2233. It is proven that the structure of aviation networks in Indonesia forms a hub and spoke system. However, if the distribution is formed following the normal distribution pattern, it shows that the structure of the flight network in Indonesia forms a point-to-point system.

It is shown in **Figure 4** that Juanda International Airport in Surabaya, Soekarno-Hatta International Airport in Jakarta, and Sultan Hasanuddin International Airport in Makassar are the airports with the highest connectivity level in Indonesia. This means that these three airports play the most important role (main hub airports) for the whole structure of aviation networks in Indonesia. Based on government regulations concerning the airport system in Indonesia as stated in Government Regulation No. 69 of 2013, the three airports are included in the primary hub airport category, a level below I Gusti Ngurah Rai International Airport in Denpasar, which is included in the category of secondary hub airport [16]. However, the level of tertiary hub airport has two other airports, namely Kualanamu International Airport in Medan and SAMS International Airport in Balikpapan. In addition to these six airports, the other airports serve as spoke airports.

6.2 Herfindahl: Hirschman index method

HHI is a distribution and market size measure equal to the sum of squares of participation of a corporation "i" ($HHI = \Sigma Pi^2$). HHI is widely implemented in transportation to analyse the relationship between ticket prices and market concentration and correlates directly with the number of the effective market participants.

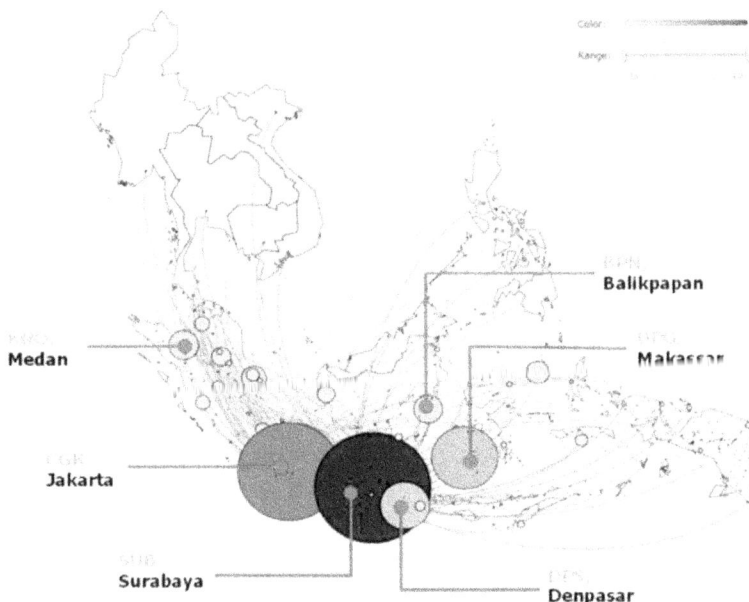

Figure 4.
Airport system based on eigenvector centrality.

No.	Airport	Number of international air passengers**	π	π^2
1	Sultan Iskandar Muda	74,380	0.00309	0.00001
2	**Kuala Namu***	2,153,244	0.08946	0.00800
3	Minangkabau	89,455	0.00372	0.00001
4	Sultan Syarif Kasim II	447,858	0.01861	0.00035
5	Hang Nadim	31,634	0.00131	0.00000
6	S.M. Badaruddin II	121,987	0.00507	0.00003
7	Husein Sastranegara	295,849	0.01229	0.00015
8	**Soekarno-Hatta***	11,849,161	0.49229	0.24235
9	Halim Perdana Kusuma	14,562	0.00060	0.00000
10	Ahmad Yani	116.426	0.00484	0.00002
11	Adi Sumarmo	124,016	0.00515	0.00003
12	Adi Sutjipto	294,701	0.01224	0.00015
13	**Juanda***	1,822,372	0.07571	0.00573
14	**Ngurah Rai***	6,140,721	0.25512	0.06509
15	Lombok	143,872	0.00598	0.00004
16	Supadio	23,101	0.00096	0.00000
17	**SAMS***	98,808	0.00411	0.00002
18	Juwata	22,008	0.00091	0.00000
19	Sam Ratulangi	46,530	0.00193	0.00000
20	**Sultan Hasanuddin***	150,445	0.00625	0.00004
21	Syamsudin Noor	8362	0.00035	0.00000
	$\sum Xi$	24,069,492		
	HHI	0.322		
	ne	4		
	n	29		
	$n^2 - ne.n$	725		
	$(n^2 - n^2.n)^{0.5}$	26.92582		
	$n - (n^2 - ne.n)^{0.5}$	2.07418		
	$0.5(n - (n^2 - ne.n))^{0.5}$	1.03709		
	Number of hub	2		

Note: *Year of 2014. **Sample of airports.

Table 2.
HHI method to determine hub airport on supporting open sky ASEAN.

The method of HHI becomes attractive because it correlates directly with the number of effective market airports, for example if the HHI in a market has a value 0.1, then this market will be as competitive as the 10 companies with the same large size. The values of HHI should be in the range of "0" to "1"; however, in the case of airports only in the range 0 to 0.5, every time an aircraft takes off one is landing at another airport.

In the HHI method, the input parameter used as the basis for determining a hub airport is the production of air transport movement (passenger, cargo). In this research, as many as 21 airports that provide international flights are analysed. The flights that are calculated are the number of international flight passengers. **Table 2** analyses the calculation of the HHI method.

In **Table 2**, the calculation result shows that the distribution of international air passenger production has an HHI value of 0.322%, meaning this value gives an indication that the distribution of international air passenger production has a high concentration on a certain airport. The data in the table shows that the airport with the biggest international air passenger production is Soekarno-Hatta Airport-Jakarta, with 11,849,161 passengers or 49.23% of international flight total movement. Further, the number of effective airports is four, as follows:

1. Soekarno-Hatta Airport-Jakarta;

2. Ngurah Rai Airport-Bali;

3. Kuala Namu Airport-Medan; and

4. Juanda Airport-Surabaya.

The four effective airports are designated as ASEAN hub airports. Sultan Hasanuddin Airport-Makassar, although established as an ASEAN hub airport, is not an effective airport, this is because domestic air passenger production is less than Soekarno-Hatta Airport-Jakarta, Ngurah Rai Airport-Bali, Kuala Namu Airport-Medan and Juanda Airport-Surabaya. However, the position of Sultan Hasanuddin, which is geographically located in the middle of Indonesia, gives it good geographical potential to be developed as an ASEAN hub in the future.

6.3 Flight path efficiency

Furthermore, calculation of the efficiency route combines several configurations, as follows:

1. Flight route efficiency with a configuration of 7 hub airports +28 spoke airports; analysis of transportation efficiency value for an aviation network system with a configuration of 7 hub airports +28 spoke airports gives a value equal to 72.09%. This value is still too far from the efficient range (49–52%).

2. Flight route efficiency with a configuration of 6 hub airports +29 spoke airports; analysis of transportation efficiency value for an aviation network system with a configuration of 6 hub airports +29 spoke airports gives a value equal to 71.81%. This value is still too far from the efficient range (49–52%).

3. Flight route efficiency with a configuration of 5 hub airports +30 spoke airports; analysis of transportation efficiency value for an aviation network system with a configuration of 5 hub airports +30 spoke airports gives a value equal to 71.15%. This value is still too far from the efficient range (49–52%). The value of transportation efficiency does not vary much with transportation efficiency in the 6 hub +29 spoke configuration. The problem with this configuration caused the hub airport that is not taken into account (SAMS Airport-Balikpapan) has the flight movement not too large production movement in the system.

4. Flight route efficiency with a configuration of 4 hub airports +31 spoke airports; analysis of transportation efficiency value on an aviation network system with a configuration of 4 hub airports +31 spoke airports gives a value

equal to 66.48%. This value is still quite far from the efficient range (49–52%); however, it is better than the previous configuration.

5. Flight route efficiency with a configuration of 3 hub airports +32 spoke airports; analysis of transportation efficiency value on an aviation network system with a configuration of 3 hub airports +32 spoke airports gives a value equal to 63.21%; this is better than the previous configuration.

These results provide an understanding of the value of flight efficiency in terms of flight network modelling but is still quite far from the optimum efficiency range, i.e. 49–52%, and shows that the flight network system with various hub and spoke models needs to be improved.

7. Conclusions

The main conclusion from the research is that the simulation of flight networks is still above the optimum efficiency range, i.e. 49–59%, meaning that the result of the configuration alternatives is dynamic depending on increasing performance. Performance could be improved if there were a high concentration value of emergence of international flight networks heading to the airport. The overall conclusion is as follows:

- Regarding the existing air transport, the production of air transport in Indonesia is distributed between 35 major airports in 34 provinces. Air passenger movements are divided by the number of active domestic flight routes.

- The hub and spoke model of the existing airport system, based on the pattern of existing air transport movement, shows that the number of active hub airports is eight:

1. Kuala Namu Airport-Medan;

2. Hang Nadim Airport-Batam;

3. Soekarno-Hatta Airport-Jakarta;

4. Adi Sucipto Airport-Yogyakarta;

5. Juanda Airport-Surabaya;

6. Ngurah Rai Airport-Bali;

7. Sultan Hasanuddin Airport-Makassar; and.

8. SAMS Airport-Balikpapan.

These eight airports are considered as the existing hub and have a number of connecting flights.

- There is an alternative to the hub and spoke model that supports ASEAN open sky. With reference to government regulation number 69 of 2013 the number of international hubs that support ASEAN open sky is five

1. Kuala Namu Airport-Medan;

2. Soekarno-Hatta-Jakarta;

3. Juanda Airport-Surabaya;

4. Ngurah Rai-Bali; and

5. Sultan Hasanuddin-Makassar.

In addition to the regulation, hub and spoke patterns also consider the production of trips by trip route. Based on the production of the trip there are three airports with considerable production and a large number of flight routes:

1. Kuala Namu Airport-Medan;

2. Soekarno-Hatta Airport-Jakarta; and.

3. Juanda Airport-Surabaya.

The hub and spoke airport model system that supports ASEAN Open Sky was developed with various configurations, as follows:

a. 7 hub and 28 spoke configuration models (considering the Indonesian Economic Infrastructure Development Acceleration Program);

b. 6 hub and 29 spoke configuration models;

c. 5 hub and 30 spoke configuration models (considering Government Regulation number 69 of 2013);

d. 4 hub and 31 spoke configuration models;

e. 3 hub and 32 spoke configuration models (considering airport production).

- Efficiency of flight routes:

a. The flight efficiency of various hub and spoke configuration models is as follows:

    i. Efficient transport value 7H + 28S = 72.09%;

    ii. Efficient transport value 6H + 29S = 71.81%;

    iii. Efficient transport value 5H + 30S = 71.15%;

    iv. Efficient transport value 4H + 31S = 66.48%;

    v. Efficient transport value 3H + 32S = 63.21%;

b. The alternative to the hub and spoke models suggests that the 3 hub configuration (Kuala Namu Airport-Medan, Soekarno-Hatta Airport-Jakarta and Juanda Airport-Surabaya) is the best configuration, although it does not reach the efficiency range of 49–52%.

c. In general, the various configurations of the number of hubs for supporting the Open Sky policy are still quite far from the flight efficiency value and need to be improved.

Acknowledgements

The author is grateful to the Centre of Research and Development for Air Transportation of the Ministry of Transportation Indonesia for encouragement in writing this chapter and the academic editor who gave valuable comments on aspects of book production.

Author details

Eny Yuliawati
The Centre of Research and Development for Air Transportation, Ministry of Transportation, Indonesia

*Address all correspondence to: enjulia_2005@yahoo.co.id

IntechOpen

References

[1] Barabasi AL. Linked: The New Science of Network. Cambridge, Massachussets: Perseus Publishing; 2002

[2] Pels E. A note on airline alliances. Journal of Air Transport Management. 2001;**7**(1):3-7

[3] Alderihi M, Cento A, Nijkamp P, Rietveld P. Network competition—The coexistence of hub and spoke and point to point systems. Journal of Air Transport Management. 2005;**1**(5):328-334

[4] Dennis NP. Airline hub operation in Europe. Journal of Transport Geography. 1994;**2**(4):219-233

[5] Haryanto IW. Studi Kasus Perencanaan Sistem dan Teknik Transportasi Udara di Indonesia. Case Study of System and Engineering Planning on Indonesia Air Transportation. Yogyakarta-Indonesia: Gadjah Mada University Press; 2015

[6] ERIA Study Team. Assessment of the Implementation of ASEAN Transport Cooperation in ASEAN Strategic Transport Plan 2011-2015. Jakarta: ASEAN Secretariat and ERIA. pp. 4-1-4-74

[7] Bowen J. The Asia Pacific airline industry: Prospects for multilateral liberalization. In: Findlay C, Chia Lin S, Singh K, editors. Asia Pacific Air Transport: Challenges and Policy Reforms. Singapore: Institute of South East Asian Studies; 1997

[8] Findlay C, Forsyth P. Air transport in the Asian Pacific region. Asia Pacific Economic Literature. 1992;**6**:1-10

[9] Air Transport Policy. Workshop on Air Transport Policy. Bangkok: ICAO Printed; 2001

[10] Li M. Air transport in ASEAN: Recent developments and implications. Journal of Air Transport Management. 1998;**4**:135-144

[11] Oum TH, Yu C. Shaping Air Transport in Asia Pacific. Aldershot: Ashgate; 2000

[12] Implementation Blueprint. Brunei Darussalam-Indonesia-Malaysia-Philippines East ASEAN Growth Area (BIMP-EAGA). Philippines: ADB printed; 2012

[13] Graham B. Geography and Air Transport. Chichester, England: Wiley; 1995

[14] O'Connell JF, Bueno OE. A study into the hub performance of emirates, Etihad airways and Qatar airways and their competitive position against the major European hubbing airlines. Journal of Air Transport Management. 2018;**69**:257-268

[15] Herfindahl–Hirschman Index, The United Stated Department of Justice. Retrieved from http://www.justice.gov/atr/herfindahl-hirschman-index

[16] Tatanan Kebandarudaraan Nasional. Peraturan Menteri Perhubungan Nomor PM 69 Tahun 2013. Kementerian Perhubungan Indonesia; 2013

Chapter 4

Tree-Network Overrun Model Associated with Pilots' Actions and Flight Operational Procedures

Michelle Carvalho Galvão Silva Pinto Bandeira,

Anderson Ribeiro Correia and Marcelo Ramos Martins

Abstract

The runway excursions are defined as the exit of an aircraft from the surface of the runway. These excursions can take place at takeoff or at landing and consist of two types of events: veer off and overrun. This last one, which occurs when the aircraft exceeds the limits at the end of the runway, is the event of interest in the current study. This chapter aims to present an accident model with a new approach in aeronautical systems, based on the tasks of the pilots related to the operational procedures necessary for the approach and landing, in order to obtain the chain of events that lead to this type of accident. Thus, the tree-network overrun model (TNO model) was proposed, unlike most traditional models, which consider only the hardware failures or which do not satisfactorily explain the interrelationship between the factors influencing the operator. The proposed model is developed in a fault tree and transformed into a Bayesian network up to the level of the basic elements. The results showed the qualitative model of the main tasks performed by the pilots and their relation to the accident. It has also been suggested how to find and estimate the probability of factors that can impact on each of the tasks.

Keywords: overrun, TNO model, fault tree, Bayesian networks, safety, aviation

1. Introduction

Around the world, the occurrence of runway excursions in commercial and general aviation is the highest ones. The International Air Transport Association (IATA) and the International Civil Aviation Organization (ICAO), through the Runway Excursion Risk Reduction Toolkit [1], define runway excursions as the exit of an aircraft from the surface of the track. These excursions might take place at takeoff or landing and consist of two types of events: veer off and overrun. For the landing, they can be described as:

- Veer off (LDVO): when there is an exit in which the aircraft exceeds the lateral limits of a runway in the landing phase.

- Overrun (LDOR): when overtaking occurs at the end of the runway during the landing phase. Event of interest of the current study.

The latest Boeing data from a survey conducted from 2006 to 2015 show that the final phase and landing phase together account for 49% of fatal accidents in the world's commercial jet fleet [2]. The number of onboard fatalities on the aircraft in these same phases of flight accounts for 47% of the total. The statistic was evaluated according to the aircraft exposure time for each of the mentioned phases (percentage of flight time estimated for 1.5-h flight). The phases of this study interest—descent, initial approach, final approach, and landing—correspond together to 59% of fatal accidents and 61% of fatalities on board.

1.1 Literature review

Most of the aviation accident statistics cited in the literature today begins with the data collected in the late 1950s and early 1960s, and it is possible to observe a marked decline in the accident rate [3]. Beginning in the 1950s, a number of research efforts was undertaken to document the precise location of aircraft accidents so that effective data safety and security planning could be obtained from the airport and its surroundings. It is noteworthy that "the airport and its neighbors" identified the location of more than 30 military and commercial aircraft accidents, which occurred outside the physical boundaries of the airport with fatal victims or injured people on soil [4]. Despite limited data, this report led to the establishment of "clear zones," which are now known as "track protection zones." Besides that, they also brought important contributions to the literature: "Air Installation Compatible Use Zone (AICUZ) Program" of the US Department of Defense served to define potential areas of accident for military aircraft, known as "accident potential zones (APZs)" [5]; "location of aircraft accidents/incidents relative to runways," compiled data on the location of accidents with commercial airplanes on the airport runway [6]; and surveys conducted by the Airline Pilots Association indicated that 5% of accidents occur in route, 15% occur in the vicinity of airports, while the remaining 80% occur on runways, overpass areas, and clear zones [7]. However, the increasing complexity in technological systems, such as aviation systems, maritime systems, air traffic control, telecommunications, nuclear plants, aerospace defense systems among others, has raised points of discussion about modes of failure and related new issues to security, such as the analysis of human factors and organizational factors in a system.

To reduce these negative effects, it has been observed that studies are being carried out with a larger number of samples (accidents or incidents). As an example, there are the accident analysis studies developed by [8–14]. As a result, it was observed that this distance differs for each type of operation, whether landing or takeoff, as well as for each type of accident, whether overrun, undershoot, or veer off. The studies previously mentioned were important to present the differences among the events on runway excursions and to report which runway conditions influence each type of the accident. They also showed that aircraft operational factors are important in the analysis of an accident. Despite the contributions mentioned, they were mainly limited to environmental factors and models based on historical data. The relationship between occurrences and human performance factors, for example, was not explained.

Many researchers have attempted to develop theories or models to describe the causes of an accident [15]. One of the earliest models of accident causes is the "Domino theory" proposed by Heinrich in the 1940s, which describes an accident as a chain of discrete events occurring in a particular temporal order [16]. This theory belongs to the class of models of sequential accidents or models based on accident events, which gave subsidies for most models of analysis of accidents introduced later [17]. These models were known to use causality methods such as: failure mode and effect analysis (FMEA), fault tree analysis (FTA), event tree analysis (ETA), and cause-consequence analysis (CCA). A large part of this approach has been strongly criticized for being based only on causal relationships among the events [18–20].

1.2 Concept of the study

Safety is generally understood as a state of the transportation system; therefore, it has a qualitative nature. In aviation, there are neither widely accepted security measures, nor is there a common agreement on the limits of the indicators that can be considered acceptable [21]. In this context, interdisciplinary research and studies are necessary to understand the complexity of sociotechnical systems [18, 20]. In addition, through a broad systemic view, one can understand the multidimensional aspects of safety, to later achieve the modeling of accidents in a more global way.

Since the middle of the last century, safety models of the technical and human parts of the systems have been introduced [17]. Further studies provided important reviews of the various existing accident models [22–26]. The latter one presents an extensive research with 121 accident models described and their applications. In [25], the authors develop quantitative indicators to assess the status of the flight team and the impact of these indicators in air traffic safety. In [22], the authors particularly review the models of accident analysis, and in [27], the author develops a model for analysis of incidents using petri net, both for air traffic. In [28], the authors present a proposal to relate human factors, abilities, organizational factors and environmental factors to the task being performed by the pilot. This application proposes several relationships between these factors. These authors based on literature and research with pilots in flight simulators to obtain the results of relationship of the factors. A summary of the major accident models identified are highlighted in **Table 1** [12, 29–36].

The most recent model presents the purpose of this study. The methods or techniques that were used in these analyzes are shown in **Table 2**. The latter table was adapted according to the categories presented in [24] to classify the methods and/or techniques used. Thus, accident models can be divided into four categories: (i) causality model, (ii) collision risk model, (iii) human error models, and (iv) third-party risk model.

The TNO model is conceptually similar to [40], which uses the same tools to develop the model's ship collision accident. These authors used fault tree to obtain

#	MAIN MODELS OF ACCIDENTS IN AVIATION	ORIGINAL OR ADAPTED
1	Reich–Mark Model [29]	Original
2	SHELL Model [30]	Adapted from [37]
3	CRM Model [31]	Original
4	HFACS Model [32]	Adapted from Swiss Cheese Model [38]
5	Flight Model [33]	Original
6	Impact Relationship Map Model (IRM) [34]	Original
7	Approach Model (ACRP 50) [12]	Original
8	CATS Model [35]	Original
9	Accident Model STAMP-HFACS [36]	Adapted from STAMP Model [39]
10	Tree-Network Overrun Model (TNO Model)	Original

Table 1.
Identification of aviation accident models.

CATEGORY	CHARACTERISTICS	METHODS OR TECNIQUES	TYPE OF ANALYSIS
CAUSALITY MODEL	The models address the risk and safety assessment of aircraft operations, particularly, failures of certain technical systems and components that result in an aircraft accident. Failures can be due to many interrelated causes (aircraft and ATC / ATM).	Fault Tree Analysis	Systems Reliability Risk analysis
		Event tree analysis	
		Bow-Tie analysis	
		Monte Carlon simulation	Systems Reliability
		Bayesian network	Systems Reliability Human Reliability
COLLISION RISK MODEL	The models cover the assessment of the risk of collision of aircraft in flight and / or on the ground.	Monte Carlo simulation	Systems Reliability
		Bayesian network	Systems Reliability Human Reliability
		Petri network	Risk analysis
HUMAN ERROR MODEL	The models address the assessment of accidents and incidents due to human error (errors associated, mainly with pilots and air traffic controllers).	HAZOP	Risk analysis
		Fuzzy Logic	Human Reliability
		HEART	
		HERA	
		Bayesian network	Systems Reliability Human Reliability
THIRD-PARTY RISK MODEL	The models address the risk assessment targeted at the airport area people, who may be affected in an aviation accident.	Matrix of Probability and Consequence	Risk analysis
		Probabilits Equations	

Table 2.
Category of accident models vs. accident investigation methods.

the main human failures related to the ship's crew tasks, and Bayesian networks (BNs) to obtain the probability of collision and the relationships between the contributing factors. Two other models similar to the proposed model are the flight model [33] and CATS model [35]. The first presents a model in Bayesian networks with a selection of contributing factors in order to obtain the probability of an aviation accident. Despite the contribution of human and organizational factors, this model does not represent the main operational procedures, nor does all the flight phases. The second one, CATS model [35], presents an aviation accident model developed by fault tree, where human failure is the only element that is obtained by Bayesian networks. This implies that the top event is static in relation to the other factors, making it impossible to obtain the contribution of this element with the accident and the possibility of the relationship between the various factors of the tree.

The objective of this chapter is to present a probabilistic accident model for the landing overrun of medium and large aircraft with the purpose of evaluating operational safety during approach and landing through the pilot-aircraft interface, considering the main operational procedures and the pilot's tasks. So that, it is possible from these elements to observe the abilities and human factors of pilots, the performance of the airline, airport infrastructure, and environmental conditions in the field of commercial aviation.

2. Development of the TNO model

The methodology presents the fault tree developed to represent the chain of events, which brings the consequences of human errors. Thus, this topic presents the development of FTA and its basic elements. Then, the FTA is transformed into a Bayesian network (BN). For each basic event, a BN is developed related to the task it is associated with, in which the performance factors will be aggregated. These factors, as well as the development of the model are presented throughout this chapter.

The methodology of this research presents four stages—familiarization, qualitative analysis, quantitative analysis, and incorporation to obtain the proposed accident model. These steps were adapted from the methodology proposed by [41] that aimed a human reliability analysis (HRA).

In the familiarization stage, besides the literature review, it was consulted the technical documentation of entities related to the sector to understand the operation and to describe the procedures of approach and landing of medium and large commercial aircraft and their flight stages, in addition to the current norms emanating from the competent organs. The following references were used: ALAR report [11], risk analysis report [8], ACRP 3 [10], TAM general operations manual [42], Flight Crew Training Manual for Aircraft Model A319, A320 and A321 [43], Flight Crew Operation Manual [44], and Flight Crew Training Manual for the 737NG [45]. In addition, fieldwork was carried out in an A-320 aircraft simulator; consultation with specialists—pilots and industry analysts—was an important point for the development of the model, showing the best coherence among the relationships between the operational procedures and the pilots' activities. Finally, the accident analysis presents the NTSB database data on the causes of accidents of the LDO type, which helped the analysis of the relationships of the elements of the proposed model. Step 2 basically presents the FTA technique and the BN method used to construct the proposed model. Step 3, in summary, concerns the population of network elements developed by the model. And, step 4 presents the results and inferences.

2.1 Fault tree in the construction of the TNO model

The fault tree analysis (FTA) technique is widely used in aerospace, nuclear, and electronic systems [46]. FTA is a quantitative technique of the type "top-down" in which the top event refers to a single event from which the intermediate events lead to component failures as well as to human actions. Logical trees can be used both for a qualitative and quantitative evaluation of the system; they employ a deductive procedure to determine the possible causes of an event of interest located at the top of the tree that may be the fault or success in the execution of a given mission. The qualitative evaluation aims at identifying the cause-effect relationship between the events that may contribute to the occurrence of the top event (of interest) as well as its logical dependencies, while the quantitative evaluation aims to determine the probability of occurrence of the same top event from the probability of occurrence of the events that make up the tree. Moreover, the final objective of a qualitative analysis of an FTA is mainly the probability of occurrence of events, in addition to obtaining the set of minimum cuts and prioritizing them according to their order. **Table 3** shows the logic gates used in the current study.

It is important to emphasize that the quantitative evaluation is deterministic and performed from the basic events, not allowing a diagnostic evaluation based on the evidence, and in both qualitative and quantitative analyzes, the basic events are considered Boolean; that is, they have only two possible states. Then, the logic of the model is represented by Boolean algebra rules, where each variable may have one of the binary values corresponding to the concepts of true (1) or false (0) [47]. If the top event is the failure of a system in the execution of a given mission, the tree is said to be faulty, and if the top event is the success of the system, the tree is said to be successful. In the latter case, it is said that the probability P of the top event will be the reliability of the system being analyzed, while in the first one, the reliability of the system will be $1-P$ (top event).

LOGICAL GATE	SYMBOL	DEFINITION
AND		The exit event only occurs if all entry events occur.
OR		The exit event occurs if at least one of the entry events occurs.
OR EXCLUSIVE		The exit event occurs if at least one, but not all, entry events occur.

Table 3.
Logical gates used in the model.

2.2 Operational procedures selected for the TNO model

The development of the proposed model followed steps in which each element was designated by a number in the FTA, symbolized in the parentheses:

i. It was highlighted the landing overrun as the top event (#1).

ii. In order for an overrun to occur, it was determined that two situations must occur simultaneously: the "unwanted state in the operation of the aircraft" (#2) and the "flight crew did not go-around the aircraft" (#16). This association is warranted by the *Flight Crew Training Manual* for the A319, A320, and A321 [43] aircrafts and *Flight Crew Training Manual* for the 737NG [45] aircraft that indicate the go-around for destabilized approach in order to avoid a runway excursion. Therefore, the connection of these factors was represented by an "E" logic gate. It is worth noting that in the BN model, this event assumed a 75% probability of overrun occurrence when both dangerous events occur, and 25% of the accident does not occur under the same conditions, according to [48]. This condition is not represented in the FTA because of its Boolean structure.

iii. The "unwanted state in the operation of the aircraft" event implies in two situations: "unwanted state in the descent" (#3) or "unwanted state in the landing" (#39). Either of these two situations makes the landing operation unwanted. This way, the logic gate "OR" was used.

iv. For the "unwanted state in the descent" event to occur (#3), two situations were observed: "undesired state in the briefing" (#4) or "unwanted state in flight management" (#17). Either of these two dangerous events can lead to an undesired state of descent.

v. The "unwanted state in the briefing" (#4) was designed in consultation with experts. This way, they obtained two dangerous events: the nonexistent briefing (#5), when the flight crew decides not to make the necessary configurations for the descent procedure, and the inadequate briefing (#8) when the flight crew performs the task but does not meet the appropriate safety conditions, classified as incomplete (#14) or incorrect (#9). For the "unwanted state in flight management" (#17), they considered three situations: "inadequate checklist" (#18), "inadequate flight control" (#25) or "inadequate final approach" (#32), all linked to a logic gate "OR." These events and their ramifications were arranged according to the consultation of the possible dangerous events with experts and are based on the description of operational safety reports [49–52]. According to the literature, the cause of the factors is linked to omission or error in action, criteria not met for stabilized approximation, inadequate monitoring, among others. Additionally, the basic events were obtained with observations in the field and consultation with specialists. According to the pilots, once an error occurs in the procedure, it is quickly detected by the flight crew. The detection of the error in some of the activities developed in the proposed model has practically a 100% chance to occur. However, the error correction action may be flawed, as represented in FTA and BN (#20, #27, #34).

vi. Finally, the event "unwanted landing state" (#39) was considered to occur when there is an "unfavorable runway" (#40) or "inadequate braking" (#41). Therefore, the connection of these factors was represented by an "OR" logic gate. Such a link was justified according to the flight simulator cockpit monitoring, where an overrun event was observed in both conditions, with the approach stabilized until the moment of landing consultation with experts also suggested the occurrence of this dangerous event. In addition, the hazardous event "inadequate braking" (#41) presents the "landing gear procedure error" (#42) and the error in the reverse procedure (#43) as basic events. In the fault tree, the designated logic gate was "OU." However, the relationship of these two events was modeled in the BN with the ratio of 80% being braking adequate when the landing gear procedure is adequate and the reverse procedure is inadequate, and 20% of braking adequate when landing gear procedure is inadequate and the reverse procedure is appropriate. This condition is not represented in the FTA because of its Boolean structure.

The framework of the model proposed in FTA is in **Figure 1**. The pilot tasks that must be analyzed in the proposed model are listed in **Table 4**. The model elements with negligible failure are chosen based on field research and consultation with experts.

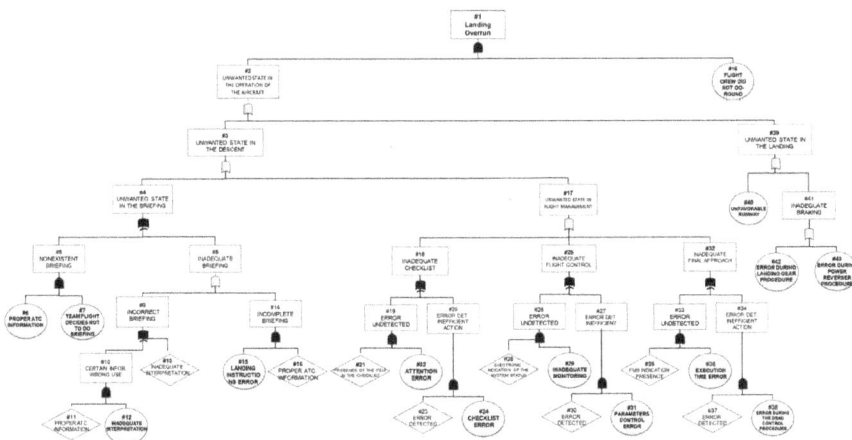

Figure 1.
Fault tree with basic events that lead to landing overrun.

i	BASIC EVENT	A	B	C	DESCRIPTION
16	Flight crew did not go-round	■			The flight crew decides, depending on the approach conditions of the aircraft and the local infrastructure, whether the landing will be carried out or discontinued.
42	Error during landing gear procedure		■		Error during landing gear procedure for not following aircraft standards and / or company standard.
43	Error during power reverser procedure		■		Error during power eversor procedure for not following aircraft standards and / or company standard.
40	Unfavorable runway			■	According to experts, the unfavorable runway is the one that is contaminated (by water, oil, ice, etc.), improper grooving (grooves that help the liquid run in the runway and increase the grip in contact with the tires) and / or one of the headers ant it is forbidden for any reason, reducing the length of the runway.
38	Error during the drag control procedure	■			Error during the drag control procedure (spoilers and speedbrakes) for not following the aircraft norms and / or standard of the company.
36	Execution time error			■	Action at the wrong time.
35	FMS indication presence		■		Flight Management System that assists the monitoring of aircraft control.
31	Parameters control error	■			Error in observing, monitoring or indicating the parameters used for speed, descent rate and power used for an adequate landing.
29	Inadequate monitoring			■	Wrong identification or unrealized observation.
23	Error detected		■		Identification of error.
28	Electronic indication of the system status		■		The status of the aircraft is indicated by the Electronic Centralized Aircraft Monitoring before its approach (for example: ECAM, for the Airbus fleet).
24	Checklist error	■			Checklist non-compliant with the company's security policy.
22	Attention error			■	Due to lack of attention to the senses, the individual makes an erroneous selection.
21	Presence of the item in the checklist		■		Appeal (printed and electronic checklist) used by the flight crew to perform the Checklist, that is, the verification of items and actions required for landing the aircraft.
15	Landing Instructions Error	■			Landing briefing not in accordance with company safety policy.
13	Inadequate ATC information			■	Information on airspace control and airport of destination that is incomplete or non-compliant with safety regulations.
12	Inadequate interpretation			■	Diagnostic failure.
11	Proper ATC Information		■		Information on airspace control and airport of destination given in accordance with safety standards.
7	Team flight decides not to do briefing			■	Error in decision making.

A: elements of model that represent the pilot task that must be analyzed;
B: elements of model with negligible failure;
C: elements of model that represent environmental and human conditions.

Table 4.
Basic elements of the fault tree (FTA).

2.3 Bayesian network in the construction of the TNO model

Bayesian network (BN) is defined as a graphical structure for representing the probabilistic relationships among a large number of variables and for making probabilistic inferences with those variables [53]. Bayesian networks—also known as opinion networks, causal networks, or graphs of dependency—are graphic reasoning models based on uncertainty that use the concept of probability as the analyst's degree of belief, allowing for expert judgments to be used as information to support a decision-making process related to complex systems [54–56]. The BNs showed to be useful in studies of system reliability [40, 57] and in risk analysis studies [58–60]. Yet, it has been applied to complex systems such as nuclear plants [61, 62], maritime transport [63, 64], and in the last 10 years, several studies on human reliability are also being developed in aviation using BNs [24, 28, 33, 35, 65–71].

A BN is a directed acyclic graph (DAG), which is defined as $G = (V,E)$, where V are the nodes representing either discrete or continuous variables and E is a set of ordered pairs of distinct elements of V, called arcs (or edges), and represents the dependencies between the nodes. The conditional probabilities associated with the variables are the quantitative components. The nodes and arcs are the qualitative components of the networks and provide a set of conditional independence assumptions, which means that each arc built from variable X to variable Y is a direct dependence, such as a cause-effect relationship and, in that case, the node representing variable X is said to be a parent node of node Y [53].

Each node within a Bayesian network is classified as "parent," "child," or both. These classifications relate to their respective relations to other nodes, where children nodes are those connected to antecedent nodes or are influenced by other nodes; parents are those connected to decedent nodes or which have an influence on other nodes [72]. Once we have specified the topology, we need to specify the conditional probability table (CPT) for each node. Each row in the table contains the conditional probability of each node value for a conditioning case. A conditioning case is just a possible combination of values for the parent nodes.

Considering a BN containing n nodes, X_1 to X_n, taken in that order, a particular value in the joint distribution is represented by $P(X_1 = x_1, X_2 = x_2, ..., X_n = x_n)$, or more compactly, $P(x_1, x_2, ..., x_n)$, and the chain rule of probability theory allows to factorize these joint probabilities as shown in Eq. (1). Then, this process is repeated, reducing each conjunctival probability to a conditional probability and a smaller conjunction, until it forms a great product as shown in Eq. (2).

$$P(x_1, x_2, ..., x_n) = P(x_n | x_1, ..., x_{n-1})P(x_1 | P(x_1, ..., x_{n-1}) \tag{1}$$

$$P(x_1, x_2, ..., x_n) = P(x_n | x_1, ..., x_{n-1})P(x_{n-1} | (x_1, ..., x_{n-2})$$
$$...P(x_1)P(x_2 | x_1) = \prod_{i=1}^{n} P(x_i, | x_1, ..., x_{i-1}) \tag{2}$$

$$P(x_1 | x_2, ..., x_n) = \prod_{i=1}^{n} P(x_i | Parents\ (X_i)) \tag{3}$$

The quantitative analysis is based on the conditional independence assumption. Considering three random variables X, Y, and Z, X is said to be conditionally independent of Y given Z, if $P(X,Y|Z) = P(X|Z)P(Y|Z)$. The joint probability distribution of a set of variables, based on conditional independence, can be factorized as shown in Eq. (3) since the constraint defined in Eq. (4) is verified. This equation allows obtaining any joint probability from values found in conditional probabilities tables, in the case of discrete variables, or from the conditional probability density function, for continuous variables. A complete example can be found in [69].

$$Parents\ (Xi) \subseteq \{X_1, ..., X_{i-1}\} \tag{4}$$

Thus, each entry in the joint is represented by the product of the appropriate elements of the conditional probability tables (CPTs) in the belief network. The CPTs therefore provide a decomposed representation of the joint. The possibility of using evidences of the system to reassess the probabilities of network events is another important feature of the BNs. Given some evidence, beliefs can be recalculated to evaluate their impact on the network nodes. The process of obtaining a posteriori probability from a priori probability is called Bayesian inference [53]. As emphasized by [73], inferences can be made using Bayesian networks in three distinct ways: causal, diagnostic, and intercausal.

2.4 Fault tree conversion in Bayesian networks

It is possible to combine a structured methodology as fault tree with the modeling and analytical power of the Bayesian network [74]. The authors also point

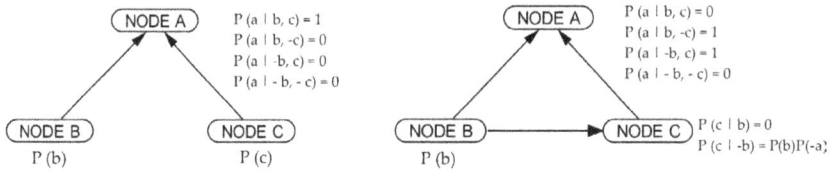

Figure 2.
BN to logic gate "E" (at the left) and the logic gate "Exclusive OR" (at the right).

out that any fault tree can be converted into a Bayesian network without losing information. It is important to note that the flexibility of Bayesian network modeling can accommodate several types of dependencies among variables that cannot be included in fault tree modeling. Studies have shown that the transformation of a problem described by a fault tree into a Bayesian network is not a complex process [74, 75]. To convert the fault tree into a Bayesian network, the basic premises of the standard FTA methodology are highlighted, as follows [74]:

- events are binary (example: appropriate/not appropriate);

- events are statistically independent;

- the relations between events and causes are represented by logic gates through Boolean logic, i.e., AND and OR gates; and

- the root of the fault tree is the unwanted event; i.e., it is the top event to be analyzed.

Thus, one node must be created for each event and for each basic element in the FTA. It is important to note that in BN, each element in the FTA must be represented only once, even if there are repetitions in the fault tree. Then, the nodes must be connected, according to the logic gates present in the FTA.

A subsystem composed of a logical gate whose Boolean algebra is of any nature (union, intersection, excluding union, or others) with k branched components, being events or subsystems, which can be converted into their corresponding Bayesian network. If the logical gate is represented by a union, then, only the nonoccurrence of all events avoids the occurrence of the top event, i.e., $(E_1^c \cap E_2^c \cap ... \cap E_k^c)$, where P $(Top|E_1^c \cap E_2^c \cap ... \cap E_k^c) = 0$, and any other combination of E_k^c leads to such occurrence. It is highlighted that according to Morgan's theorem (propositions for simplifying expressions in Boolean algebra), x^c indicates the complementary event of X, where $(X \cup Y \cup Z)C = Xc \cap Yc \cap Zc$ e $(X \cap Y \cap Z)C = Xc \cup Yc \cup$. Considering that the logic gate represents an intersection, only the simultaneous occurrence of all events leads to the top event, that is, only P $(Top|E1 \cap E2 \cap ... \cap Ek)$ is not null, being it equal to 1. **Figure 2** illustrates the conversion of FTA into BN. Each of the figures has two independent basic events A and B and the top event C.

3. Results

The result of the FTA transformation in BN is presented, qualitatively, in **Figure 3**. The Bayesian network of the proposed model presents two states, negative and positive, for each node. The negative state represents the probability of occurrence

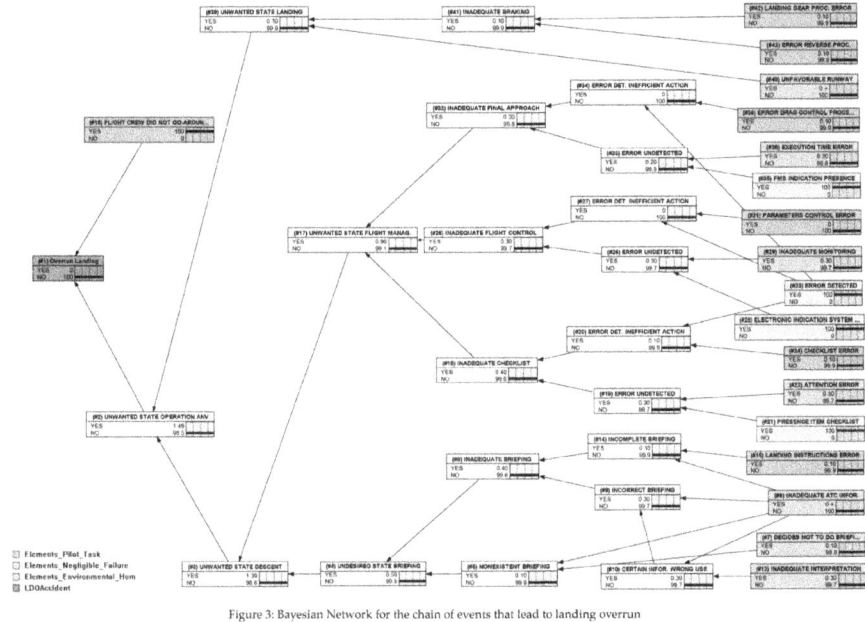

Figure 3: Bayesian Network for the chain of events that lead to landing overrun

Figure 3.
Bayesian network for the chain of events that lead to landing overrun.

of the node (characterized by the word YES). And the positive state represents the probability of not occurring the node; that is, the fault does not occur (characterized by the word NO). The node in red represents the landing overrun, and its positive and negative states represent the probability of the accident occurring, given the factors of the developed network.

According to field research and expert opinion, the tasks that require most pilots during approach and landing are listed below. On these tasks, a chain of dangerous events was also obtained, described in the development of the model as below:

- decide if the aircraft continues to approach and/or landing (go-around);

- landing briefing;

- landing checklist;

- control of aircraft parameters;

- execution of the procedure drag on final approach; and

- execution of the braking procedure (landing gear and reverse).

It should be noted that this work is not intended to introduce the factors of each task and its probabilities in this example, but to present the accident model for the approach and landing phases related to the tasks performed by the pilots that can be visually understood. The TNO model includes the main tasks performed by the crew and the chain of dangerous events that can lead to landing overrun.

From this model, it is possible to obtain the relation between the factors that can influence the performance of the pilots, and therefore, this can indicate how this can impact in the success or failure of the tasks related to the procedures of approach and landing. For each of these tasks, it is possible to develop more focused studies and to obtain the organizational, environmental, human factors, and the main abilities around each one of them. One way to get the factors contributing to the negative state of each of these tasks was suggested in [28]. Once obtained, a way to develop the Bayesian network with these factors and to find the probability of each of the states, positive and negative, is in [71].

The main advantage of transforming the FTA model to BN is to verify the sensitivity of each of the nodes given the accident and to obtain their impact. It is also possible to obtain the probability of an accident occurring because an error occurred in some task, for example. This type of approach is only possible in BN, one of the advantages of using this method for risk analysis. Finally, the network data can be obtained by consulting specialists and/or obtained from the literature.

4. Conclusions

Human factors are the most important source of uncertainties of any model, though many techniques and computational tools arise in recent decades to deal with the complexity of sociotechnical systems. To be able to get a representative analysis of the real system, a systemic vision of process is required. However, to model operational procedures of a system, or its main tasks, is not an easy step. So, first it is important to know the system that is intended to be modeled, and then analyze the factors (and their relationships) that can contribute to an occurrence. For such information, a search in the literature and a research with pilots and accident investigators become extremely important.

The proposed model was described and used to model the relationship between the main operational procedures performed by the flight crew and the pilots' skills and to support the construction of a BN to quantitatively analyze the event of interest. Differently from other studies, the TNO model proposes a systematic and efficient way to organize the influence factors through an FTA and, consequently, to obtain a probabilistic analysis through a BN. The use of BN to find the most probable cause with the objective of identifying the most important factors and prioritizing the mitigation action is also an important contribution of this work. As far as we know, no other study has proposed a similar approach.

It should be noted that factors related to component failures in aircraft systems are not being considered in the general model. This is because studies of failures in aeronautical equipment are already traditionally considered and modeled, besides presenting a low probability of occurrence. Therefore, the emphasis was placed on the human actions of pilots. Thus, our intention was to model one of the main tasks of the flight team considering factors that precede team error. This model must be able to obtain a representative analysis of the real system; a systemic view of the process is also needed. In this sense, this model of accident fills this gap.

The results indicate subsidies to propose mitigating actions and can collaborate with the management of air transport operational safety. The best way to improve the latter is to attack the most sensitive points. Thus, the factors highlighted in the analysis, once prioritized within the company, can promote the reduction of runway excursions during the landing procedure of medium and large aircraft.

Acknowledgements

The authors are especially grateful to the pilots and engineers who participated in the consultations conducted by the research. The authors are grateful for the financial support given by CAPES and FAPESP in Brazil for this study.

Author details

Michelle Carvalho Galvão Silva Pinto Bandeira[1*], Anderson Ribeiro Correia[1] and Marcelo Ramos Martins[2]

1 Department of Aviation and Airports, Technological Institute of Aeronautics, São José dos Campos, SP, Brazil

2 Naval Architecture and Ocean Engineering Department, University de Sao Paulo, São Paulo, SP, Brazil

*Address all correspondence to: mgalvao@ita.br

IntechOpen

References

[1] IATA, ICAO. Runway Excursion Risk Reduction Toolkit. 2nd ed. Montréal, Québec: IATA & ICAO; 2012

[2] Boeing. Statistical Summary of Commercial Jet Airplane Accidents: Worldwide Operations 1959-2015. Seattle, Washington: Boeing Commercial Airplanes; 2016

[3] Wiegmann DA, Shappell SA. A Human Error Approach to Aviation Accident Analysis: The Human Factors Analysis and Classification System. England: Ashgate Publ; 2012

[4] President's Airport Commission. The Airport and Its Neighbors. Washington, DC: Government Printing Office; 1952. 116p. Available from: http://www.dot. state.mn.us/aero/planning/documents/ airportanditsneighbors.pdf [Accessed: June 3, 2012]

[5] U.S. Environmental Protection Agency. EPA 550/9-77-353: Air Installations Compatible Use Zones (AICUZ) Program. Washington, D.C.: Federal Noise Program Reports Series; 1977;**1**:96

[6] Federal Aviation Administration. Location of Commercial Aircraft Accidents/Incidents Relative to Runways. Washington, DC: FAA; 1990. Report No. DOT/FAA/AOV90-1

[7] Ashford N, Wright PH. Airport Engineering. 3rd ed. New York: Wiley-Interscience Publ.; 1992

[8] Eddowes M, Hancox J, Macinnes A. Final Report on the Risk Analysis in Support of Aerodrome Design Rules: A Report Produced for the Norwegian Civil Aviation. United Kingdom: AEA Technology; 2001. Report No. AEAT/ RAIR/RD02325/R/002

[9] CALTRANS. California Airport Land Use Planning Handbook. California: California Department of Transportation; 2002. 455p. Available from: http://www.dot.ca.gov/hq/ planning/aeronaut/documents/alucp/ AirportLandUsePlanningHandbook.pdf [Accessed: May 06, 2012]

[10] Hall J et al. ACRP Report 3: Analysis of Aircraft Overruns and Undershoots for Runway Safety Areas. Washington, DC: Transportation Research Board, TRB; 2008

[11] Flight Safety Foundation. Runway Excursion. Alexandria, VA: ALAR; 2009. Available from: http://www. skybrary.aero/bookshelf/books/865.pdf [Accessed: March 28, 2015]

[12] Ayres M et al. ACRP Report 50: Improved Model for Risk Assessment of Runway Safety Areas. Washington, DC: Transportation Research Board; 2011

[13] European Aviation Safety Agency (EASA). Annual Safety Review. Germany: EASA; Available from: https://www.easa.europa.eu/sites/ default/files/dfu/218639_EASA_ASR_ MAIN_REPORT_2018.pdf [Accessed: November 29th, 2018]

[14] International Air Transport Association (IATA). Safety Report. Montreal: IATA; 2014. Available from: http://www.iata.org/publications/ Pages/safety_report.aspx. [Accessed: January 12, 2015]

[15] Roelen ALC. Causal risk model of air transport - comparison of user needs and model capabilities. PhD thesis. Delft University of Technology, 2008

[16] Heinrich HW, Petersen D, Roos N. Industrial Accident Prevention: Safety Management Approach. 5th ed. New York: McGraw-Hill; 1980

[17] Leveson N. Safeware: System Safety and Computers. 1st ed. New York:

Addison-Wesley Professional Company, Inc.; 1995. 704 p. ISBN: 0-201-11972-2

[18] Rasmussen J. Risk management in a dynamic society: A modelling problem. Safety Science. 1997;**27**(2):183-213

[19] Hollnagel E. Barriers and Accident Prevention. Hampshire: Ashgate Publ; 2004

[20] Leveson N. A new accident model for engineering safer systems. Safety Science. 2004;**42**:237-270

[21] Skorupski J. About the need of a new look at safety as a goal and constraint in air traffic management. Procedia Engineering. 2017;**187**:117-123. DOI: 10.1016/j.proeng.2017.04.357

[22] Machol RE. Thirty years of modelling midair collisions. Interfaces. 1995;**25**:151-172

[23] Greenberg RA. Quantitative safety model of systems subject to low probability high consequence accidents [Thesis]. Univ. of South Australia; 2007. 465f

[24] Netjasov F, Janic M. A review of research on risk and safety modelling in civil aviation. Journal of Air Transport Management. 2008;**14**(4):213-220

[25] Skorupski J, Wiktorowski M. Chapter 129: The model of a pilot competency as a factor influencing the safety of air traffic. In: Nowakowski T, editor. Safety and Reliability: Methodology and Applications. London: Taylor & Francis Group; 2015. DOI. 10.1201/b17399-138

[26] Hughes BP et al. A review of models relevant to road safety. Accident Analysis and Prevention. 2015;**74**:250-270

[27] Skorupski J. Modelling of traffic incidents in transport. TransNav: The International Journal on Marine Navigation and Safety of Sea Transportation. 2012;**6**(3):357-365

[28] Bandeira MCGSP, Correia AR, Martins MR. General model analysis of aeronautical accidents involving human and organizational factors. Journal of Air Transport Management. 2018;**69**:137-146

[29] Reich P. Analysis of long range air traffic systems: Separation standards—I, II and III. Journal of the Institute of Navigation. 1966;**19**:88-96, 169-176, 31-338

[30] Hawkins FH. Human Factors in Flight. 2nd ed. Aldershot, England: Avebury Technical; 1993

[31] Helmreich RL, Merritt AC. Culture at Work in Aviation and Medicine: National, Organizational, and Professional Influences. England, Brookfield, VT: Ashgate, Aldershot; 1998

[32] Shapell S, Wiegmann D. The Human Factors Analysis and Classification System (HFACS). Washington, DC: Federal Aviation Administration; 2000. Report No. DOT/FAA/AM-00/7

[33] Greenberg R, Cook SC, Harris D. A civil aviation safety assessment model using a Bayesian belief network (BBN). Aeronautical Journal. 2005;**109**(1101):557-568. DOI: 10.1017/S0001924000000981

[34] LIOU JJH, Yen L, Tzeng G-II. Building an effective safety management system for airlines. Journal of Air Transport Management. 2008;**14**(1):20-26

[35] Roelen ALC, Lin PH, Hale AR. Accident models and organizational factors in air transport: The need for multi-method models. Safety Science. 2011;**49**:5-10. DOI: 10.1016/j.ssci.2010.01.022

[36] Lower M, Magott J, Skorupski J. A system-theoretic accident model and process with human factors analysis and classification system taxonomy. Safety Science. 2018;**110**(Part A):393-410. DOI: 10.1016/j.ssci.2018.04.015

[37] Edwards E. Man and machine: Systems for safety. In: Proceedings of British Airline Pilots Associations Technical Symposium. London: British Airline Pilots Association; 1972. pp. 21-36

[38] Reason J. Human Error. Cambridge UK: Cambridge University Press; 1990

[39] Leveson N. Engineering a Safer World, Systems Thinking Applied to Safety. London: MIT Press; 2012

[40] Martins MR, Maturana MC. Application of Bayesian belief networks to the human reliability analysis of an oil tanker operation focusing on collision accidents. Reliability Engineering and System Safety. 2013;**110**:89-109

[41] Swain AD, Guttmann HE. Handbook of Human Reliability Analysis with Emphasis on Nuclear Power Plant Applications. Albuquerque, NM, USA: Sandia National Laboratories; 1983. Report No. NUREG/CR-1278-F, SAND80-0200

[42] TAM. Manual Geral de Operações (MGO). Rev. 05. São Paulo, SP: Airlines TAM; 2011. 752 p. Available from: https://pt.scribd.com/document/246034510/MGO-TAM-REV5-pdf [Accessed: June 19, 2016]

[43] AIRBUS. Flight Crew Training Manual: Reference Tam A319/A320/A321.[S.l]. Blagnac Cedex, France: FLEET FCTM; 2013. 412 p

[44] AIRBUS. Flight Crew Operating Manual: Reference Tam A319/A320/A321. [S.l]. Blagnac Cedex, France: FLEET FCOM; 2014. 618 p

[45] Boeing. B737 NG/MAX: Flight Crew Training Manual. Revision Number 14. Washington: Boeing; 2015. 404p

[46] Lee WS et al. Fault tree analysis, methods, and application: A review. IEEE Transactions on Reliability. 1995;**34**(3):194-203

[47] Modarres M. What every Engineer Should Know about Reliability and Risk Analysis. New York: Marcel Drekker, Inc.; 1993

[48] National Transportation Safety Board (NTSB). Aviation Accident Data Summary. Accident Number: DCA13FA1312013. Washinigton, DC. 2015. p. 12. Available from: https://app.ntsb.gov/pdfgenerator/ReportGeneratorFile.ashx?EventID=20130723X13256&AKey=1&RType=Final&IType=FA [Accessed: February 16, 2016]

[49] Flight Safety Foundation (FSF). Approach and Landing Accident Reduction (ALAR). Alexandria, Virginia: FSF ALAR Task Force; 1998. Available from: http://flightsafety.org/current-safetyinitiatives/approach-and-landing-accident-reduction-alar [Accessed: March 28, 2015]

[50] Airbus. Flight Operations Briefing Notes (No. FLT OPS – SOP – SEQ 01 – REV 04), Standard Operating Procedures (SOPs): Operating Philosophy. Toulouse, France: Airbus Customer Services; 2006

[51] International Federation Of Airline Pilots Associations (IFALPA). Runway Safety Manual. Montreal, Québéc: IFALPA; 2009

[52] Federal Aviation Administration. Instrument Approach Handbook. Washington, DC: FAA; 2014. Available from: http://www.faa.gov/regulations_policies/handbooks_manuals/aviation/instrument_procedures_handbook/media/Chapter_4.pdf [Accessed: May 08, 2015]

[53] Neapolitan RE. Learning Bayesian Networks. New Jersey: Pearson P. Hall; 2004

[54] Pearl J. Probabilistic Reasoning in Intelligent Systems. San Mateo, California: Morgan Kaufmann; 1988

[55] Jensen FV, Nielsen TD. Bayesian Networks and Decision Graphs, Information Science and Statistics. New York, NY: Springer; 2007

[56] Cowell RG, Dawid P, Lauritzen S, Spiegelhalter D. Probabilistic Networks and Expert Systems: Exact Computational Methods for Bayesian Networks, Statistics for Engineering and Information Science Series. New York: Springer; 2007

[57] Schleder AM, Martins MR, Modarres M. The use of Bayesian networks in reliability analysis of the LNG regasification system on a FSRU under different scenarios. In: Twenty-Second International Offshore and Polar Engineering Conference. Rhodes, Grécia: ISOPE; June 17-22, 2012. pp. 881-888

[58] Ale B, Van Gulijk C, Hanea A, Hanea D, Hudson P, Lin PH, et al. Towards BBN based risk modelling of process plants. Safety Science. 2014;**69**:48-56. DOI: 10.1016/j.ssci.2013.12.007

[59] Martins MR, Schleder AM, Droguett EL. A methodology for risk analysis based on hybrid Bayesian networks: Application to the regasification system of liquefied natural gas onboard a floating storage and regasification unit. Risk Analysis. 2014;**34**(12):2098-2120

[60] Martins MR et al. Quantitative risk analysis of loading and offloading liquefied natural gas (LNG) on a floating storage and regasification unit (FSRU). Journal of Loss Prevention in the Process Industries. 2016;**43**:629-653

[61] Sundaramurthi R, Smidts C. Human reliability modeling for the next generation system code. Annals of Nuclear Energy. 2013;**52**:137-156

[62] Martins MR, Maturana MC, Frutuoso PFF. Methodology for system reliability analysis during the conceptual phase of complex system design considering human factors. In: International Topical Meeting on Probabilistic Safety Assessment and Analysis. Sun Valley, ID: NAS/PSA; April 26-30, 2015. pp. 1-14

[63] Martins MR, Maturana MC. Human error contribution in collision and grounding of oil tankers. Risk Analysis. 2010;**30**(4):674-698

[64] Zhang G, Thai VV. Expert elicitation and Bayesian network modeling for shipping accidents: A literature review. Safety Science. 2016;**87**:53-62. DOI: 10.1016/j.ssci.2016.03.019

[65] Ale BJM, Bellamy LJ, Cooke RM, Goossens LHJ, Hale AR, Roelen ALC, et al. Towards a causal model for air transport safety—An ongoing research project. Safety Science. 2006;**44**:657-673. DOI: 10.1016/j.ssci.2006.02.002

[66] Luxhøj JT, Coit DW. Modeling low probability/high consequence events: an aviation safety risk model. In: Reliability and Maintainability Symposium. Newport Beach, CA, USA: IEEE; June 14-16, 2006. pp. 215-221

[67] Ale BJM et al. Further development of a causal model for air transport safety (CATS): Building the mathematical heart. Reliability Engineering and System Safety. 2009;**94**(9):1433-1441

[68] Mohaghegh Z, Kazemi R, Mosleh A. Incorporating organizational factors into probabilistic risk assessment (PRA) of complex socio-technical systems: A hybrid technique formalization. Reliability Engineering and System Safety. 2009;**94**:1000-1018. DOI: 10.1016/j.ress.2008.11.006

[69] Brooker P. Experts, Bayesian belief networks, rare events and aviation risk estimates. Safety Science. 2011;**49**:1142-1155. DOI: 10.1016/j.ssci.2011.03.006

[70] Bandeira MCGSP, Correia AR, Martins MR. Method for measuring factors that affect the performance of pilots. Transport. 2017;**25**(2):156-169

[71] Bandeira MCGSP, Correia AR, Martins MR. Landing accident model for medium and large sized commercial aircraft. In: 22nd Air Transport Research Society World Conference, Seoul, South Korea; 2018

[72] Stanton N, Landry S, Di Bucchianico G, Vallicelli A. Advances in human aspects of transportation: Part III, Advances in human factors and ergonomics. In: AHFE Conference; 2014

[73] Russell J, Norvig P. Artificial Intelligence: A Modern Approach. Upper Saddle River, N.J: Prentice Hall/Pearson Education; 2010

[74] Bobbio A et al. Improving the analysis of dependable systems by mapping fault trees into Bayesian networks. Reliability Engineering and System Safety. 2001;**71**(3):249-260

[75] Hamada M et al. A fully Bayesian approach for combining multilevel failure information in fault tree quantification and optimal follow-on resource allocation. Reliability Engineering and System Safety. 2004;**86**(3):297-305

Aviation of the Future: What Needs to Change to Get Aviation Fit for the Twenty-First Century

Ursula Silling

Abstract

The world around us has changed dramatically, particularly since the beginning of the twenty-first century, mainly due to the broad availability of the Internet. Inventions such as smart phones, apps, virtual face to face conversations, coupled with the rise of Facebook, Google, Amazon & Co. added a lot of speed to this development. The digital revolution empowers the consumer and determines ever increasing expectations. At the same time, latest tech developments such as artificial intelligence (AI), machine learning (ML), blockchain, voice and more create opportunities never seen before. However, the aviation industry to a large extent has remained stuck in legacy processes and their decades old technology. It also suffers from low profit margins. With a few exceptions, aviation management overall struggles on how to adapt to the real-time and agile environment. Digital transformation activities have started both in operational and commercial areas, but fundamental underlying platforms and culture change in most cases have not yet been addressed. This chapter explains reasons behind key pain points of the industry, what activities are ongoing and the main areas that need to change to get into shape for the current dynamic environment.

Keywords: digital transformation, change management, legacy processes, technology, agile, leadership, artificial intelligence, blockchain, customer experience, aviation, airline, airports, travel agencies, tour operators, airlines, airports, modern management, multi-speed IT, distribution, digitisation, sales, travel retail, technology, machine learning, data, digital cockpit, digital airport, digital airline, amazon of the air, travel retail, business model, strategy, aircraft on demand, travel tech

1. Introduction

The target of this chapter is to provide some glimpse behind the curtains, some results of empirical and cross industry research as well as my personal observations and experiences over time. I will focus on why the aviation industry has been slow to adopt the changes, give more background about the underlying problems and outline what activities are already happening and which are the four key opportunities which absolutely need to be tackled. This is not meant to be a complete list of what is happening in the industry, but rather about some of the game changers and critical success factors to bring about change, based on our extensive experience and insight over the years as well as ongoing market research. I am not tackling sustainability, even though I think it is a key problem that needs to be addressed separately, not just in terms of the impact of fuel consumption but also in terms of the amount of plastic created during each flight, the airport operation and impact on the environment, and the problems of over-tourism.

Let me start with an illustration of the as is situation by pointing out some of my recent travel experiences. In June 2018 I travelled from Switzerland to the US as I was a speaker and judge at a big travel tech event in San Francisco. During the flight I had to use internet as I still needed to send an urgent email. When I asked the flight attendant why internet was not working she shrugged her shoulders and said she did not know. 2 h later I tried again and finally managed to send my email. At arrival in San Francisco the queues for passport control were so long that people could not get off the running walkways. It took more than 2 h to get out of the airport. I had to continue my trip to Asia before going back home. I tried several times to book a ticket directly with a large Asian carrier, but I could not complete the process as the payment options did not foresee any European credit cards. I was forced to book with an online agency instead, and their booking process did not allow me to book a seat. Lost seat revenue and higher ticket cost because of agency commission are what this meant for the airline. For me it meant a lot of wasted time and frustration. On the last part of my flight back to Switzerland, a woman from Chicago sitting next to me was crying as she had lost her previous connection and had been running so hard to get on this flight—as missing it would have meant an overnight—that she had left her laptop bag in the aircraft. The airline crew at the gate was very unfriendly with her, and she felt completely helpless. She was visiting her boyfriend in Switzerland for only a couple of days, and while the super friendly flight attendant had already been able to tell her that her luggage had been found, it only finally arrived at her boy-friend's address several days after she had already been back home in the US. It took her several phone calls and being stuck in waiting lines to contact centres to get there.

These experiences contrast sharply with a world where I write invoices with my mobile phone, buy products and services at Amazon and Alibaba with one click, switch off the light at my home by talking to Alexa, answer my doorbell even when travelling thanks to the smart doorbell Ding, order a present for my Mum online and let it be delivered to my car's boot, get flight status updates by just entering a flight number in google, order my dinner for my late flight via app, for pick up at the airport restaurants or even gate delivery. Where do these visible problems come from in an industry which in its early days had so much pride in customer service and innovation?

2. The state of the industry: and why flying can be so painful

The aviation market has always been quite volatile. Even going back to regulated environments airlines have gone from a wave of positive results to huge losses. They have been extremely exposed to external factors, from new legal restrictions to fuel price change and politic and economic impact on demand for air travel. Airports as being even more capital intensive have seen their performance as a consequence of airline decisions. The rise of the low cost carriers was not taken seriously initially by the full service traditional network carriers before they reached significant market share and started to enter the lucrative long haul sector as well.

2.1 Airline profitability

For the aviation industry dependence on external factors such as fuel, labour cost, the political environment and economic growth factors has always been extremely high. The Gulf war illustrated this very clearly, as did the rise of low cost carriers in the 1990s, 11/9/2001 and the global economic crisis starting in 2008. These events led airlines to rethink their aircraft ownership or lease strategies as well as increased focus on their cost structures. Ryanair as a game changer for the European and global airline market had turned to the low cost model when facing huge losses and realising that they could only survive with drastic change. They questioned everything they did, aligned processes and product proposition and seized the opportunities which the broad availability of internet provided in terms of efficiency and customer reach without the necessity of large investments into sales infrastructure. They started to reinvent themselves again a couple of years ago with the introduction of significant customer service improvements "... and begin to manage those customers and deliver individually tailored service for them to meet their needs" [1], when realising the limits their model had reached.

The subsequent global growth of low cost carriers can be attributed to extreme cost focus and subsequent large price differentials to traditional carriers, frequency of service, flexibility to abandon routes if they do not perform, the rise in economic activity and increased internet penetration and e-literacy, increase in purchasing power of middle class households particularly in developing regions, ease of travel, urbanisation and changes in lifestyle and consumer preferences with the widespread availability of the smart phone and the control that the internet rendered to consumers. While many attempts at long haul low cost operations had failed, there has been some radical change in recent years, with Norwegian Airlines being one of the key drivers, attacking the main profit makers of the traditional network carriers.

The latter had already started in the 1990s to found their own low cost carrier. Yet as they did not let them develop completely independently they often failed and incurred extremely high losses as their cost structures and behaviour was too much aligned with what the airline group did. Lufthansa's subsidiary Eurowings is one example. Go by British Airways was sold to Easy Jet and latest attempts include long haul low cost with their subsidiary Level as a reaction to Norwegian Airlines' growth in the lucrative long haul market. Emirates is moving to an alignment of network and customer proposition such as their frequent flyer program with their low cost subsidiary flydubai after they had originally been independent. There are still more recent low cost carrier start ups by network carriers, for example Swoop, West Jet's new ultra-low cost carrier and flyadeal, Saudia Airlines' new low cost subsidiary.

In recent years, traditional airlines started to unbundle their service offering and followed what low cost carriers had been doing as part of their strategy: they added price tags for luggage, early boarding, hold fees and more. The interesting thing is that this happened in a period when the low cost carriers reached more maturity and started to enhance their customer proposition and to target business travellers with tailored services. This leads to the somehow paradox situation that network carriers still claim to offer more service, yet factually customers can choose their way of flying for much lower fares and not rarely better service with low cost carriers.

Low fuel rates, relatively high growth in demand for air travel (7–8% versus a 20-year average of 5.5%), growing seat load factors and the adoption of more and more ancillary services for sale helped to achieve a positive performance again for airlines in the last years. In some regions such as the US the intensive consolidation has also helped to increase average fares and thus total revenue. International Air Traffic Association (IATA) announced in June 2018 that it expects airlines to achieve a collective net profit of $33.8 billion, with a net margin of 4.1% in 2018 [2].

This result is driven to a large extent by North American airlines, followed by Asia-Pacific and European ones.

However, this is a downward revision from the previous forecast and compares to US$38 billion in 2017, mainly driven by increase in cost of fuel, labour and interest rates.

According to IATA [2], airfares keep going down. The 2018 average return airfare (before surcharges and tax) is expected to be US$380 (2018 dollars), which is 59% below 1998 levels after adjusting for inflation. Average air freight rates for 2018 are expected to be US$1.80/kg, which is 63% below 1998 levels.

An analysis of the Forbes Global 2000 list [3] gives some interesting insights in terms of financial perspective, particularly market capitalisation. Looking at the top 10, there is not a single airline or airport part of it. Yet for the first time since 2015, China and the US split the top 10 evenly this year. On the inaugural list in 2003, there were just 43 companies from the Greater Chinese Area. Meanwhile, Japan, the United Kingdom and South Korea also broke into the top five countries with the most companies.

In comparison to C-trip (which also owns Skyscanner) and Expedia, most airlines market capitalisation is in the best case close or much lower. In comparison to tech companies the gap is simply enormous. This is illustrated clearly in **Figure 1**. The one airline which does stand out is Delta, which is with US$37.1 billion in a much better position to the other airlines, with the next best one being American Airlines followed by IAG. Delta's CEO Ed Bastian [4] has realised the role of technology as a competitive advantage-next to the people in the airline- and invests heavily. When adding airports to the list, it is interesting that Aeropuertos Españoles y Navegación Aérea (AENA) seems to come close to Amadeus' market capitalisation, while all the others are significantly lower.

If you compare airline value with some of its IT providers, then you realise that Amadeus as a key IT provider to airlines is worth more in terms of market capitalisation than the airlines Lufthansa, IAG/British Airways, Air France-KLM and SAS

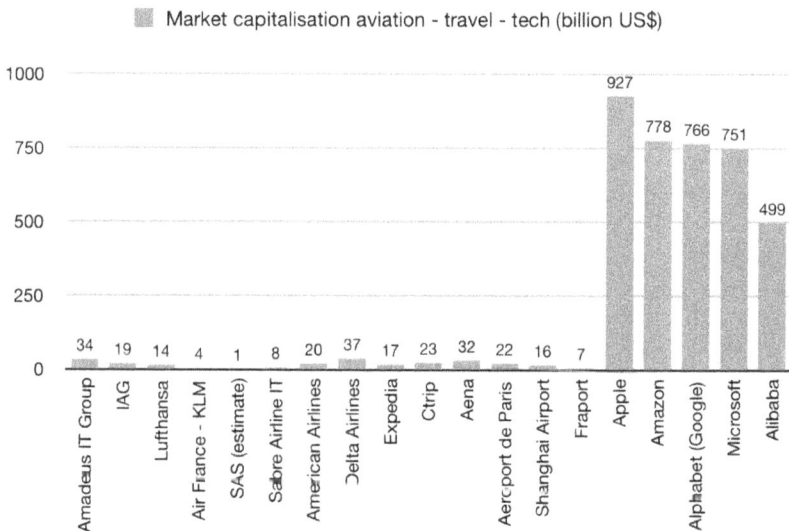

Figure 1.
Market capitalisation (US$ billion), extract 1—market capitalisation aviation and travel companies versus travel tech and technology companies overall. Data from Forbes Global 2000 [3], illustration by the author.

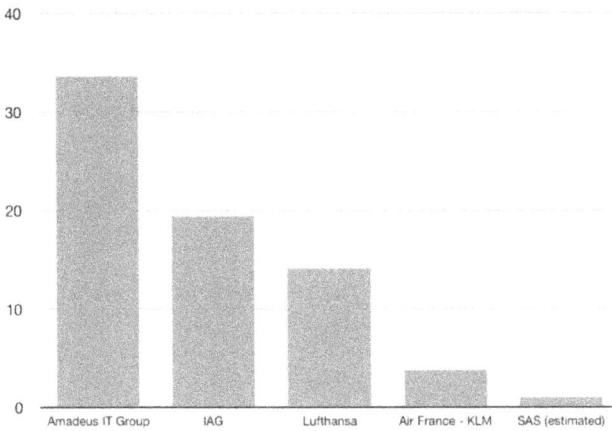

Figure 2.
Market capitalisation (US$ billion), extract 2—Amadeus versus its founding airlines.

that originally founded it 30 years ago (**Figure 2**). In fact, decades ago airlines had been very innovative and developed their own IT to be able to handle reservations and the underlying operational requirements. American Airlines had founded Sabre, Delta had founded Worldspan, Lufthansa had founded Lufthansa Systems. Many more airlines globally had developed their own IT systems in the 1950s and 1960s predominantly.

Sabre, the equivalent provider of airline solutions to Amadeus that was founded by American Airlines in 1960 is estimated to have a market capitalisation between US$7 and US$8 billion. This compares to an estimated US$19.9 billion for American Airlines [3], thus in this case the IT provider representing less than half of the value than the airline which had founded it.

For the complete picture it should be mentioned that the big traditional airline solutions providers Amadeus, Sabre and Travelsky have also a vested interest in the travel agency market by providing the Global Distribution System (GDS). They incentivise travel agencies to use their systems while they charge airlines for those distribution services [5].

2.2 Airline technology and processes

Given low profit margins and focus on operational issues and safety first, airlines in most cases simply have not had the money to invest in state of the art technology. But it is also—if not even more—the lack of priority of technology for top management. Most aviation leadership teams have been set up with more traditional management, where digital and also customer centricity had been underestimated and misrepresented. It takes a long time to change this mindset even when bringing in additional individual talent to adjust.

Airlines are used to iterative and process thinking, to a great degree influenced by legal frameworks to ensure a safe operation, but also by the decade old systems being in place and very much an inward looking culture. Top management had not realised the importance of digital. Ecommerce was evolving in a separate department with some specialists but had not really become part of the overall strategy until recently. The mindset of the workforce is significantly influenced by this process thinking approach, traditional leadership and the complexity and barriers of the current systems landscape.

Airline and airport staff often do not know why they do things. They just do it because it has always been done that way. And because their environment does not encourage questions. This leads to a number of pain points which get completely absurd in the current environment. Let me just give a few obvious questions as examples:

- Why do I need to check in? If I buy a cinema ticket or goods in the store, I pay and I get what I paid for without further validation

- Price levels for flights are restricted by numbers of letters of the alphabet instead of true commercial requirements

- Why can I not dynamically adjust change fees, e.g., by period ahead of booking, colour of shoes you are wearing, day on which you are making the changes

- Why can I not book luggage for me just for the return flight, a meal for my husband and priority boarding and a seat free next to her for my Mum

- Why do I get offered seats at check in even though I have already booked them

- I paid much for my seat, yet short term aircraft changes might mean I cannot get the seat anymore which I had reserved

- Why should airlines still spend time and money to load prices via the Airline Tariff Publishing Company (ATPCO)

- Why can I not book add-ons/ancillaries if I had booked the flight with tour operators

- Why are the additional services I had bought (seat, luggage, car) not changed as well when I make flight changes

- Why do I still receive these tickets with long text and lots of abbreviations

- Why can codeshare partners offer lower fares on the operating carrier flights than the operating carrier itself

- Why do airlines need codeshares when I could connect directly with the other airlines, which is also more transparent for customers

- Why do I not get offered more services by my airline for the airport & destination

- Why can I not start my booking on one device and continue on the other

- Why do I not just get the possibility to use the next available low cost flight if a network carrier cancels a flight ad hoc

- Why are there still cabins in the plane: one customer might look after the best seat to sleep, the other might want a good meal, etc.

- Why do I need to wait at the luggage carousel and the queue at the lost luggage desk when it is already known that my luggage was left at the departure location

- Why are data all over the place and not easily accessible nor comparable, making it very difficult for airline staff to really help to solve issues but results in fragmented processes

- Why do I not have one view of the customer but only data referring to specific flights

- Why do accounting systems have a different truth to other systems

- Why do revenue management systems still focus mainly on historic data and do not include real time information

- Why is it so costly and takes so long to make system changes, often inhibiting both certain commercial activities as well as realisation of service improvements and innovations

This list of pain points is just an extract. The pain points cover all parts of the customer journey, from trip planning to booking, experiencing and sharing. They are a result of continuing with processes and systems which had been created for a different environment, where internet did not exist and in which the technological possibilities were more limited.

The traditional systems landscape is extremely fragmented and complex, and many of the new elements such as the online channel, optional services for sale, mobile, self service for customers and staff, reporting, customer notifications had to be added on top of it as workarounds (**Figure 3**). And the traditional processes around this are still to a great degree manual and broken, and based on specialist

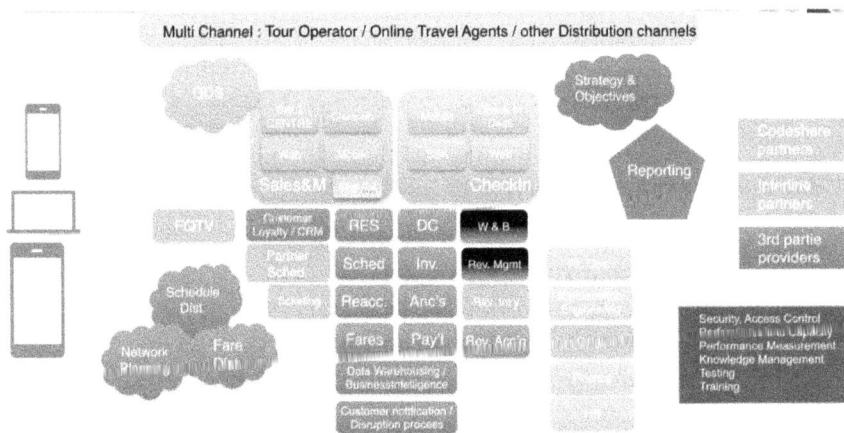

Figure 3.
Typical airline IT architecture. Abbreviations used: GDS global distribution system, FQTV fare quote system, schedule distribution, fare distribution, RES reservations, Sched scheduling, Reacc. Reaccommodation, Inv. inventory, Anc's ancillary services, Pay't payment, W&B Weight and balance, Rev.Mgmt. Revenue management, RevInt'y revenue integrity, Rev. Acc'g revenue accounting, Flt ops flight operations, HR human resources, API application programming Interface, AI artificial intelligence, ML machine learning, IOT internet of things, AR/VR stand for augmented reality and virtual reality.

silos instead of a holistic approach. They were focused on transactions and had not put the customer in the centre nor did they target a seamless experience or have foreseen the commercial and competitive pressures that we encounter today [6].

There were a number of computer failures and outages in recent years and months, from Delta, Southwest Airlines, United to British Airways [7]. Part of the underlying reason are complexity both of systems and processes, with a large degree of legacy technology, and subsequent problems to find the error. The impact is even higher as manual or alternative processes are often not in case, leading to huge disruption for customers and the airline as a result. The underlying principles and processes had been standardised via IATA initiatives, in order to make cooperation between airlines and airports and travel globally easier. IATA in recent years took a number of initiatives to adjust them to better fit with the current age. Yet it is difficult to turn around a tanker, and these are small steps in comparison to what we would expect as normal in the current digital environment.

Technology spend by airlines and airports are estimated to have reached nearly US$33 billion in 2017 [8]. This is almost exactly the total market capitalisation of Amadeus IT Systems alone. According to reference [8], top of the agenda for both airports and airlines are cyber security, cloud services and passenger self-service. Airlines' expenditure as a percentage of revenue was about 3.3% in 2017. For airports, the figure was about 5% for this year or US$8.43 billion. For 2018, it is expected that at least the same levels are being maintained, if not increased. These investment figures do not seem huge given the digital agenda but rather look like maintaining status quo. While new technology makes it possible to take a smarter approach with much less money than aviation is used to, it first requires the investment in the change. Hidden in the average figures there are airlines such as Delta and Ryanair, which are investing heavily, while a large part just work on maintaining status quo and do the most urgent adjustments. Given the high amount of investment over time and the amount of people employed coupled with resistance to change, there are lots of economic interests by providers and some other stakeholders to maintain the status quo as long as possible. In a time when the only thing which is clear about the future is that flexibility is required, providers are still trying to achieve 10 or 15 years contracts and to even restrict some commercial flexibility with regard to distribution policies. There are first signs that some big airlines do not accept this anymore, with a particular breakthrough by Lufthansa and their introduction of a distribution fee (see also Section 3.3) (**Figures 1–4**).

3. What is being done: a selection of initiatives

We had some vivid discussions and lots of examples of current activities during our last annual global think tank "think future - Hamburg Aviation Conference", bringing together top leaders, innovators and thought leaders from airlines, airports, rail, hospitality, other travel stakeholders, innovative travel tech and universities to discuss solutions how to succeed in the current dynamic environment. The live stream for this year's event can be watched on YouTube [6]. We particularly recommend the opening panel discussion between top leaders from airports, airlines and tech providers and the panel about the future for airports as additional insight to the following sections, directly from aviation thought leaders.

I think the good news is that in the meantime even the most traditional airlines' and airports' boards and executive teams have realised that change is not a choice anymore. But they struggle with the how, and what to focus on. I have summarised a few activities by airlines and airports and selective other travel stakeholders, which I think give a good impression in terms of what initiatives are prioritised out there at the moment.

3.1 Structural technology changes

Delta Airlines- having declared technology as a key focus has brought in-house two key technology platforms, its reservations and passenger services system and its flight operations systems. These are old systems they have already been using, so there was no migration required. They bought the rights from their provider Travelport. Delta had owned Worldspan—which then became part of Travelport—back in the early days of the airline. By controlling these systems Delta hopes to not only be able to act faster but also to be able to develop one view on the customer. Virgin announced in August 2018 that they will launch a new loyalty scheme with Delta in an attempt to offer a joint scheme for their customers [9].

A few low cost airlines developed their own distribution and passenger services systems (PSS) to be able to achieve best possible flexibility, Easy Jet and Jet2 in the UK being key examples.

Some airlines have decided to choose one of the more recent players in the area of underlying reservations and operations systems—to note that "recent" is relative to the majority of the systems in use today, it still means systems which were founded more than 20 years ago—such as Radixx (founded 1993), Bravo Passenger Solutions (founded 1993) and the most recent one IBS Software Services (founded 1997). ITA software, which had started to develop a completely new Passenger Services System (PSS), was bought by Google for US$ 700 million in 2010 as a vehicle for Google to further develop their travel capabilities. Since then, Google developed many features including Google flight searcher, directly linking to the relative airlines.

A number of older airlines which still own their own PSS systems—for example Aer Lingus, Iberia and Air New Zealand—are evaluating change to an external provider. The fact that this has not happened is a good example that some of them do not think that just moving to one of the existing external provider swill solve their issues. The IAG group is an illustrative for this: British Airways uses Amadeus, but Aer Lingus and Iberia both still own their own internal system.

A number of airlines have also started to think about what some of the more modular systems and add on processes such as revenue management and network planning as well as group management and operations planning of the future should ideally look like given the changed environment.

3.2 Customer experience improvements and revenue increases

Customer self service activities have been a priority for a couple of years. It is now increasingly extended to other areas such as self connecting and additional servicing via chatbots. All types of airlines started to offer additional optional services, and also charge for them, particularly for seats and luggage and other ancillary services. Yet often this has been more of a panic activity to recover poor revenue results, and the experience is often not completely thought through, with failures in terms of luggage and seat delivery by the traditional airlines in particular as they own a diversity of aircraft types. The bundling of services is an attempt to facilitate the sales process, often determined by technology restrictions as well.

Latest attempts focus on data analysis and one view of the customer in order to be able to sell more personalised products and services. In addition, beyond the pre-departure and inflight services there is more focus on the airport and destination experience. The following selection is a result of our ongoing research.

Delta tackled the luggage delivery issues in 2016 and invested US$50 million in technology so that travellers will be able to track their luggage via an app, from the moment they check their bags to the minute the bags arrive at their destination. For 2018, they focus on re-organising all their customer related data to achieve one view of the customer [10].

JetBlue has invested in Gladly through its venture arm, JetBlue Technology Ventures. Gladly is the maker of a customer service platform for various companies, including airlines, helping to achieve a customer centric service with one view of the customer.

Ryanair has started a project declared to become the Amazon of the air, as part of their "always getting better campaign". As part of this initiative they have created a customer login—which has been in place with Easy Jet and other airlines for many years already—and keep adding optional service offerings related to travel [11].

IATA has initiated a number of projects to support the airline industry—particularly New Distribution Capability (NDC) and One Order to achieve a better view on the customer and enable sales of ancillary products regardless of which distribution channel is used.

KLM focuses on social media as a way to enhance customer service, but even as a sales channel. This initiative came about during the ash cloud, when they realised the difficulties of communicating with their customers via the limited contact centre channels, as a result of which many customers approached them via social media. They are strong with their social media proposition both in Europe but also in their key regions, adapting to local preferences such as we chat in Asia. However, they also realised that the actual operational delivery is lacking behind and announced recently that they have just launched a project and released significant budget to focus on this [12].

Lufthansa and United Airlines recently declared the development of a new digital services platform (DSP) [13] that will further align the Star Alliance carriers. So far, the travel experience for customers is still fragmented, in particular in terms of additional services such as seat reservation and luggage bookings. For example, they launched a seat selection feature in June 2018 which allows a United Airlines customer to select a seat on Singapore Airlines flights booked via united.com or the United App. It means that a customer can now select a complimentary seat for the entire journey at time of registration regardless of which Star Alliance carrier is involved. At the moment this is just possible at check in.

Airlines have started to introduce digital concierge services by using multilingual chatbot technology. Finnair and Sun Express are just a few of the airlines realising this as a way forward for better customer service around the clock and increased efficiency. It focuses so far mainly on information related to bookings, but booking services are in the making as well as adding voice. But it requires a process alignment first in order to add real value.

Seat resale and upgrade offering products that airlines such as LATAM have started to introduce are more examples how airlines can solve some operational problems due to overbooking and improve the customer experience as well as gain additional revenues.

Moscow Domodedovo airport turned itself into a shopping mall, thus attracting additional visitors and revenues. Many airports had traditionally only focused on the b2b customers. But in the meantime they have realised that they do need to get better customer insight and to keep up with customer expectations. Airports such as Copenhagen, Heathrow and Dublin have introduced customer programs, in an attempt to allow for sales of additional services, customer insight and direct communication with the customer. Many airports have introduced services such as fast track and airport parking for sale online or via an app. Geneva Airport and most UK and Italian airports are examples. Also pre-order and pick up at arrival of duty free products has become a common feature. Yet it is still difficult to find exactly the retail offering at the airport ahead of your trip. But more recently this is being extended to include all the retailers at the airport, and even in town, with pick up at the airport, via an online sales offering for customers. The German company AOE have started to offer these services via their digital platform at Auckland and Frankfurt. Heathrow Airport have just announced that they will join.

Grab is an innovative company which allows to pre-order food at the airport and grab it on your way to the gate [14]. Their solution is already integrated in a number of airport apps or websites, for example London Gatwick and Heathrow Airport adopted this offering. As airport food and beverage offering have improved significantly this could become a solution to the poor quality yet high cost for the airline of offering food during flights. American Airlines and a couple of other airlines have already decided to include this offer in their customer proposition. Airlines just need to have the open mind to test this as a complete solution for food on board. Hamburg Airport has just introduced a test for preorder and delivery of breakfast at the gate, thus saving valuable time in the morning for their customers.

Amsterdam Airport and Hamburg Airport tested in 2018 improvements for the customer experience through the PASSME [15] project, which uses technology and some airport design elements to reduce the unwanted travel time and helping to spend their time according to their preferences. Tampa airport introduced a program to get more customer insight and build an action plan for higher customer satisfaction, making use of technology to support the process. Incheon/Seoul Airport have extremely efficient biometric identification at security control, which speaks to the customers in the language of their passport.

Other travel stakeholders have also done an enormous amount of customer experience improvements. Transport for London created a unified API to allow a more seamless travel experience for customers [16]. The German rail operator Deutsche Bahn improved their customers' experience by turning the DB navigator into a travel concierge, allowing clients' time to be spent effectively and according to their priorities instead of wasting it with travel planning [17].

Expedia have adapted a completely agile approach in terms of testing which websites and costumer propositions work best. They also experiment with Voice by developing a number of solutions for Alexa by Amazon [18].

Kayak and Expedia have all started using chatbots that can learn what consumers like and deliver appropriate suggestions for travel products to buy. American Express just bought Mezi, which is a personalised travel assistant based on AI supporting business travel agencies to offer multiple services for their customers, including "please just buy the same flowers as every week".

3.3 Efficiency increases

Low cost or hybrid carriers such has Virgin Express and later Brussels Airlines had already worked with surcharges for more expensive channels more than 10 years ago. These were relatively small carriers in the global context and therefore did not create much awareness or subsequent change.

In 2015, Lufthansa announced a 16-euro surcharge [19] on each booking made through global distribution systems (GDSs) like Amadeus and Sabre. Other carriers such as British Airways and Air France followed. They want customers to book directly through their websites to be able to get a better customer understanding, control their experience, offer ancillary services for sale and introduce more flexible pricing as well as ad hoc offers at the airport as for example lounge access. And they aim to control the high direct and indirect cost created through GDS bookings.

Airlines and airports are increasing the focus on self service. This leads to the increased availability and push of self service luggage check in, as Air New Zealand and Lufthansa have had in place at their home airports for a couple of years. Self connecting services to simplify connecting traffic and enable connections with low cost carriers have started to take ground since Easy Jet announced cooperation with long haul carries such has Norwegian and West Jet [20], and Air Asia introduced a special product for this.

In Japan airports are testing robots to carry heavy luggage and to clean airport premises. Munich Airport in cooperation with Lufthansa is also running a pilot to test Pepper, the humanoid robot to answer customer questions at the airport [21]. Fraport introduced the "Smart Data Lab", in an attempt to gain useful knowledge and insights and be able to take action from the data in the organisation.

3.4 Organisation design to incorporate digital, retail and innovation

Some changes in terms of realising the importance of digital and innovation have become visible in the organisational setup, both in terms of new functions and an increased presence in the top leadership. Titles such as Customer Experience Director, Digital Transformation Officers [22], Digital Officers and Innovation Officers or Directors have become quite common. Dependent on the stage of the organisation, digital is often still seen as an add-on, which becomes visible in titles such as "digital customer experience" and / or separate functions for ancillary services and loyalty instead of taking a holistic approach. "Retail" has become part of the nomenclature in organisations in some airlines and is already very common in airport organisations. Some organisations, in an attempt to stress the customer focus, have also renamed operational areas, for example "airport customer delivery" instead of "ground handling".

However, the main base of the organisation is still very similar to what it used to be, even though the functions and activities should change as they are not really aligned anymore with the current world. Revenue Management & Pricing for example is becoming increasingly mingled with digital channel pricing and sales, ancillaries and loyalty services overlap, digital channel experience and customer experience overall overlap and so on.

Throughout my career I have noticed that aviation companies often prefer re-organisation instead of tackling the key problems of revising processes to be fit for the future, assigning and building the right talent and departments working in silos.

3.5 The rise of innovation labs

Both airports and airlines have started to take initiatives to foster innovation via innovation labs.

To name a few real life examples from the airline world:

- Easy Jet puts disruptive thinking at the heart of its digital strategy and invested in Founders Factory [23].

- Ryanair established Ryanair Labs as an internal solution as part of its "always getting better" campaign.

- Lufthansa created the Lufthansa Innovation Hub as a separate subsidiary.

- IAG, in partnership with L. Marks, launched the Hangar 51 program in 2016 to help improve airport processes, digitise business processes, improve data driven decision making to enhance customer satisfaction and to develop completely new innovative ideas that can make a difference to customers.

- Jet Blue created a venture arm to foster innovation, Jet Blue Technology Ventures.

- Malaysia Airlines has launched its first in-house innovation lab last year. It is called iSpace. Malaysian claim that the opening marks the third phase of its digital transformation. Tata Consultancy Services, IBM Bluemix, Amadeus, Telekom Malaysia and University of Malaya are partnering with the airline in the initiative.

But also airports are taking attempts to innovate and support digital transformation. There is a lot of potential through digitisation to speed up and increase efficiency for processes and to develop new experiences:

- Manchester Airport Group have launched its own technology and e-commerce business to respond to technology-driven changes in the way passengers travel. They want to move the airport experience into the digital age.

- Group ADP (Paris Airports) launched the "Smart Airport" innovation hub initiative to design the airport of tomorrow.

- Munich Airport has recently announced the development of a future focused innovation campus.

- San Diego International Airport's Innovation Lab is a collaborative environment where companies, innovators and airport executives work together to create and test new ideas. The aim is to drive airport innovation and improve the customer experience. Successful ideas have the opportunity to be implemented at San Diego, other airports, and even in other relevant industries like malls, hotels, convention centres, etc.

Made by many, a digital innovative agency in London has done research on innovation labs, with a broad collection of best practice knowledge ([24], see also **Figure 4**).

They look at four main experiments related to innovation labs: the impact of proper design, the impact of actual competition, the impact of hard targets and the impact of tranquillity. The report reveals plenty of valuable insights and data, about where the blockers to innovation are, what innovation lab talent looks like (and how to manage it), how to integrate with the sponsor organisation, and why innovation labs are to business what science-fiction is to literature. Above all, and perhaps most valuably, made by many defined the key reasons why innovation labs fail, and what critical success factors are. **Figure 4** is the summary of the key learnings from the report.

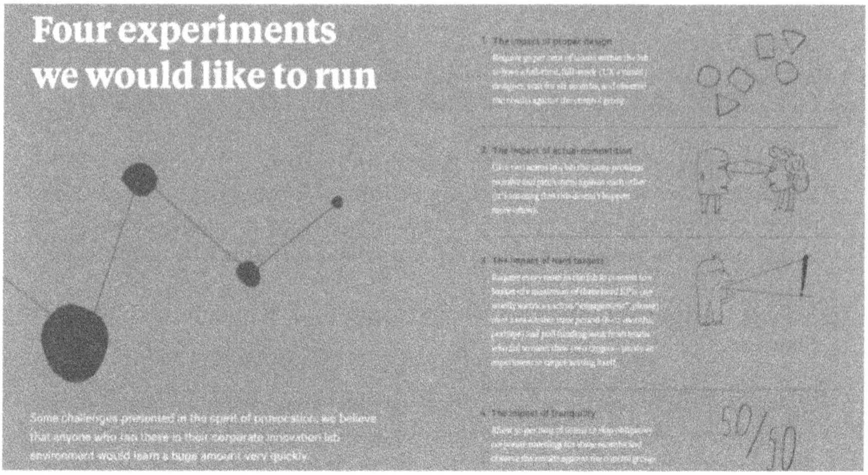

Figure 4.
Made by many, Kevin Braddock: Innovation labs—best practice, main conclusions; madebymany.com

IATA have started to support aviation by running hackathons to develop innovative solutions based on the IATA standards such as New Distribution Capability (NDC). These hackathons help to show what can be done to achieve the culture change so much needed in the industry. Unfortunately airlines are not yet making enough use of these possibilities.

3.6 Innovative things in the making: newcomers, innovating and disrupting

New technology and fresh thinking can help significantly to challenge and improve the current way of working, current profitability models, customer and staff experience, operational and commercial areas. It would be beyond the scope of this chapter to go further into details, but just showing some of the revolutionary developments in the market gives an idea of the possibilities.

A number of solutions help to overcome the silos within organisations and also foster more open thinking with external partners. In particular airports and airlines have missed a lot of opportunities because of building frontiers around themselves and not cooperating closely.

We outline a number of innovations from travel tech start ups and enabling technologies that reflect new thinking—not only new technology for old ways of doing things.

3.6.1 Augmenting customer experience and making travel planning easier

- Group Travel digitise the process of group bookings, reducing manual work and allowing to include a lot of additional services, facilitating the cooperation within the organization and between tour operators and airlines

- Trvl Porter: a style concierge recommending wardrobe for travellers to rent and making it available at their destination, no need to carry luggage any more

- AirPortr offers the service in London to pick up/deliver your luggage from/to your home or hotel and check it in for your flight

- Bounce is a start up allowing travellers to store their luggage with hotels and retailers whenever needed

- kiwi.com—helps to find all kinds of flights and develop a journey including low cost and full service airlines as well as other means of transport; they operate a contact centre as well to support customers in case of any operational disruptions

- TrustaBit uses blockchain technology to allow airlines to automate the compensation process, including the possibility to distribute vouchers during disruptions at the airport

- A number of inter-modular solutions such as Rome2Rio evolve, for airlines these solutions and new technologies make it much easier and less complex and costly than today to partner with other airlines, local taxi companies, and even boat taxis or bicycle rentals in order to get travellers exactly where they want to go and how they want to go there

- Boni Loud Steps developed indoor navigation for the visually impaired

- Interes is an innovative retail engine which helps airlines to develop and control dynamic product and promotional approaches adapted to their target groups, with pricing with no limits of the traditional systems related to fare filing or letters of the alphabet

- Hopper predict future price evolution and advise customers when best to book; they also offer alternatives to the destination chosen in line with customer preferences and budgets

- Grab allow mobile (pre)ordering for retail products and services at airports

3.6.2 Faster, more efficient, more revenue

- Automated aircraft checks conducted by robots and AI will speed up the turnaround process considerable, helping airlines to plan more efficiently

- New technology, such as 3D printing, offer new aircraft and engine design opportunities

- Data can be used to anticipate customer numbers in order to reduce crew requirements and engine maintenance, allocating the most suitable aircraft, or the most suitable gate at the airport even in case of delays. This allows more efficient staff planning. Beontra is one of the companies which developed models for integrated capacity, traffic and revenue planning to already achieve this in terms of airport planning

- Winding Tree is a start up allowing safe direct transaction with third parties by using blockchain technology, this can also help to foster the airport—airline cooperation

- YieldIn is a revenue management solution making it possible to align business priorities and revenue management practices, thus overcoming silos and ensuring engagement by top management

3.6.3 Safer and/or more sustainable and eco-friendly

- Helmets are being developed that include an augmented reality (AG) display. Pilots will be able to track all of the controls, alerts, signals, etc. more easily. Training will become more immersive as well as a result.

- The solution via Trvl Porter to "rent" your clothes at the destination saves fuel and thus is a more sustainable solution than carrying luggage.

- Further enhancements for "self flying" using AI and Machine Learning are in the making.

3.6.4 Substitutes on the horizon for current aviation models and processes

- What if Google, Amazon, Alibaba do move forward even more into travel and re-invent the whole model?

- Amazon had made some advances into travel some years ago [25] and stopped the initiative, yet technology has advanced even more now and they might give it another go given their expertise in online frictionless retail and 300 mn customer base [26].

- Alibaba has already shown significant muscle to play a major role in the Chinese travel market in spite of a strong player such as Ctrip. With their investment in a new brand Fliggy based on their Alitrip infrastructure they target the younger digital generation and have created a kind of travel market-place, allowing travel players to create their own shop while providing market-ing and data analytics support for airlines and travel players. If they combine this even more with their retail expertise and innovation activities this could potentially become a game changer.

- Google keeps adding elements of the travel journey, linking already to a number of airlines directly via Google flight search and adding travel partners to Google Maps; could they become the GDS of the future?

- Could there be completely new players in the market? What if there was just a market place for retail services and modular web based services to resolve inventory, wiping out a lot of the current processes?

- What if the principles of easy flying - which we still tend to call low cost services - becomes the norm for both long-haul and short-haul travel?

- If check in was eliminated, what would the large check in areas in airport ter-minals be used for? Could the stores just become mobile and move around the airport - where the customers are instead of directing customers to the stores? Will the order of food & beverage turn into delivery at gate services via robotics?

- Waves as a model for "flying on demand" is a start up which does already oper-ate in the UK.

- Electric and hybrid engines and models will support new models such as "fly-ing cars" and revised Concordes.

- Hyperloop as an alternative to longer distance travel.

4. What still needs to be done for the industry to survive

As seen just with the selection in Section 3 there are a number of activities ongoing in aviation to adjust to the digital age. Are they really the right things? Are they enough?

From an external view a lot of these activities seem to be little things just to get to the "normal" standard of today, and it is hard to understand why they take so much effort. And real structural issues seem to be missing. If your house is dump, just adding some high quality paint on top of the dump walls will not help. If you drive a vintage car, you will not normally use it to drive on the motorway, unless there is an emergency and you know you will be driving far too slow.

Digital innovation by Google, Amazon, Facebook, Apple, Samsung, Alibaba and other tech players but even other travel players such as online travel agents and meta search companies Expedia, Skyscanner/Ctrip and various new start ups has been out pacing the rate of change in aviation for several years, and the speed is accelerating, putting airlines and airports at a disadvantage to other industries and even to other travel stakeholders. The Forbes 2000 [3] list examples from Section 2 and the profitability and market capitalisation figures are a clear result of this (**Figure 1**).

Potential substitutes as described in Section 3.6.4 could become a real threat or simply a driver for faster and more drastic change. Coming back to the house example, it is as if avoiding to go to the basement because you know that it is full of water and old wiring and fragile walls, but you restore your house above and ignore this, hoping you will be able to continue as long as possible.

More drastic change is needed than copying current business models such as ancillary revenues or putting more focus on the customer and adding technology workarounds to make this happen. But only a few airlines and airports are really serious about it, starting to go down into the basement.

Sir Tim Clark, Emirates Airline president expressed a warning recently in an interview with Business Insider. "Guys, there's a storm coming, and if you don't get on it and deal with it, you will perish," Clark said in a recent interview with Business Insider. "The company of the 2050s will bear no resemblance to the company of 2018."

"It's not a question about using advanced technology to increase the way you do your business, like ancillary revenue streams, because that's a given," Clark said emphatically. "It's not a question of not knocking your companies down internally and rebuilding them on digital platforms. That's a given for us. It's not the case for a lot [of other airlines]." [27] Tim Clark made a major change by hiring a high calibre Chief Digital Innovation and Transformation Officer into his team end of 2016.

I believe there are 4 key areas which need to be tackled more seriously to really create a sustainable future for aviation. The model with the 4 Bs that we created is not iterative or a once off thing to do, but is meant to be re-applied on an ongoing basis, referring back and forth between the different stages and continuously evolving (**Figure 5**).

4.1 Big vision

The activities that airlines and airports currently perform are in most cases not part of a holistic strategy. They do add certain capabilities, without questioning enough the current processes and set up. If you see the tremendous amount of change happening outside of the industry, it is certain that consumer expectations will increase even more significantly.

More drastic change is needed

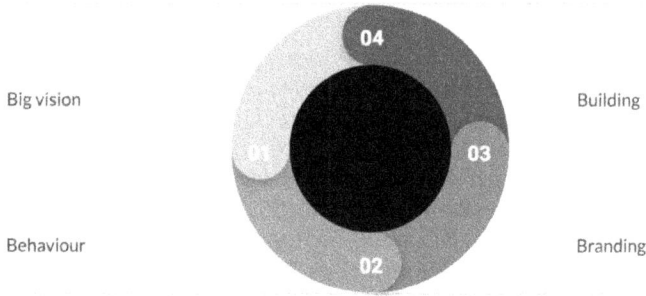

Big vision

Building

04

03

01

02

Behaviour

Branding

Figure 5.
The 4 Bs model, XXL Solutions.

Digitisation and technology based on digital platforms are a must, not even part of any vision any more.

A big and bold vision, starting with "greenfield" thinking and how you would set up an airline/airport without considering current processes. Only subsequently you would decide which of the current processes to eliminate completely, which ones to improve, which factors to build on and enhance in order to get closer to your vision.

Some airlines such as Ryanair have claimed they want to become the Amazon of the air. But Amazon has been continuously re-inventing itself, and is again doing so now with their Amazon Go stores, moving forward into the food supply chain and the internet of things (IOT) with Amazon Echo.

For airlines, their retailing ambitions so far are mainly based on adding ancillary services and optimising revenue and using more data analysis. And it seems each one is just following the others. Yet decade old technology and manual processes in distribution, revenue management and even operational areas will not provide the flexibility anymore to be ready to adapt. Unfortunately there is not the one technology solution out there to choose from, which delivers all the possibilities and flexibility needed today. But there are all the technological opportunities to implement a vision, without the complexities and large investments needed in the past.

4.2 Behaviour and mindset

A complete makeover is needed—including sorting out the basement of the house, or building a completely new house.

To develop the big vision and the subsequent makeover strategy requires above all the right leadership and mindset. And a lot of energy and care. When Ryanair re-invented the way of doing business in the 1990s their biggest risk was to have to close down. They questioned everything and used the opportunities of technology. When Willie Walsh turned ailing Aer Lingus into a low cost carrier at the beginning of this century he put very bold targets in order to achieve change and the thinking of what is needed to get there, even though it seemed far away and impossible at the time. West Jet as one of the global carriers with double digit profit margins for years as a very charismatic leader at the helm.

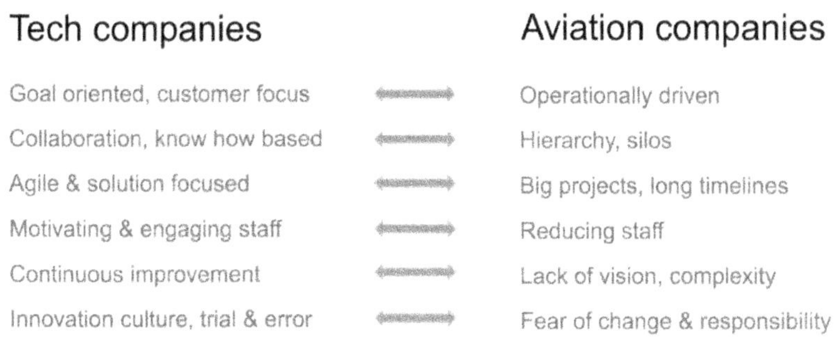

Tech companies		Aviation companies
Goal oriented, customer focus	⟷	Operationally driven
Collaboration, know how based	⟷	Hierarchy, silos
Agile & solution focused	⟷	Big projects, long timelines
Motivating & engaging staff	⟷	Reducing staff
Continuous improvement	⟷	Lack of vision, complexity
Innovation culture, trial & error	⟷	Fear of change & responsibility

Figure 6.
XXL solutions, what aviation can learn from technology companies.

The big opportunity is that technology today allows to do everything we want to—it just requires a smart approach and a big vision to get there, which in turn requires the right behaviour and mindset.

If you look at how successful technology companies are, it helps to step back and think a bit about how they work and what aviation can learn from this. In **Figure 6**, we have pointed out some of the main relevant differences between the two types of businesses. Even though they are not always completely evolved, the tendencies in terms of behaviour are very relevant.

We believe that a change of behaviour and mindset is crucial for airlines and airports to achieve any change. The current prevailing iterative and process oriented style is counter productive for the dynamic and agile digital environment. Developing an agile approach, collaboration and using best talent related to a project rather than the one who should be there according to hierarchical thinking is a key element of success for tech companies but not yet for aviation. Trust and personal responsibility are at the heart of this behaviour.

The following elements can help to create this behaviour:

A. **People and talent** are key. It requires a full review of the talent required to meet the digital and innovation requirements. You can only think out of the box if you have different boxes. Bringing in some younger people (for example by working with universities, for recruitment of new jobs or ad hoc activities) and creating more diversity to ensure more out of the box thinking coupled with training and support for existing staff to support change can help to speed up the digital transformation process. It is important to ensure that these people are really involved and can value. I have often seen some really good talent being left aside because of organisational dynamics. Travel brings together people with all kinds of different lifestyles and cultural background, yet airports and airlines are still very much national/local staff apart from the flying staff. It has also become very evident when looking at the picture of the airline CEOs at the last IATA Annual General Meeting (AGM) in Sydney that there was only one woman present.

B. **Visible changes** such as collaboration tools like Slack, Facebook for work and introducing methodologies such as design thinking and Kanban can help to support the cultural change and to break down silos. Creating time, for example Fridays or 3 h per week for innovation, making use of co-working office space can help to grow with lower cost and foster an innovation and change culture and open minds. Even innovation awards—with simple rewards such as having lunch with the CEO could make the focus very visible.

C. **Events and thought leadership from external sources,** or which training to do to look out of the current boundaries and comfort zones can help to develop the open minded behaviour needed. Aviation tends to shut down even more in case of high result pressure. Many airlines had initiated a stop on travel activities in recent years because of result pressure. In my opinion this is exactly the opposite of what should be done in the current environment. Disruptive events such as hackathons, travel tech start up events, tech and retail trends events and innovative think tanks should be used strategically to help board members, executive team members and other staff members think out of the box and develop their agenda for success. Even inviting thought leaders to do a presentation about trends in the marketplace and what it will look like in the future is something which can add to out of the box thinking for the whole company. Also Coursera, itunesU, Udemy or edX offer opportunities for staff development and training which did not exist before—and had traditionally been extremely expensive and not personalised.

D. **Define standards how to work**. This should include the principle of collaboration, agile behaviour, fast results, allowing trial and error, and to ensure to choose state of the art suppliers and do not just exclude them because they do not yet have enough customers. We are in an environment now where there is no fixed roadmap but a lot of possibilities, but it needs to be determined by the aviation stakeholders. Truly innovative suppliers and partners can help to foster innovation and open mindedness.

E. **Putting yourselves into your customers shoes.** I have seen it too often that they had no clue what happens at the airport, or on the website, as they booked in a different way and were never asked to talk to customers or make observations at different touch points of the customer journey.

Obviously, being close to customers and understanding their mindset is one of the keys to ensure the right behaviour. If customer satisfaction and feedback become part of the board meeting agenda and the executive team meetings, and technology—which is already available—is used to analyse it and immediately direct it to the right people to take action, then this already key to realise current and future needs to satisfy customers. Tesco, the UK retailer, have also created household panels and feedback opportunities at specific touch points to really get a 360 degree of customers and realise trends early.

4.3 Branding and selling

Branding and selling is meant to be both for internal and external purposes to drive change and help staff to engage and fully understand what their role is, and to retain existing customers and create awareness for new ones. It is what companies often forget; they start building a new house first and create fears as staff see the preparatory works. Ryanair's "always getting better" campaign and the vision to become Amazon of the air is a good example of initiating a major change process to reinvent themselves and establish a new market positioning.

Their fast activities and visible results are the "moments of truth" and significantly help to make people belief in the change.

Change is always linked to fears, in particular in this fast changing environment. Fears of losing their jobs because of functional changes or introduction of AI should be foreseen and rather thought about ahead. Addressing those fears, defining where human intervention can add value and start foreseeing these changes can make a significant difference. If staff are taken seriously and get engaged they can play a key role to turn around the company and establish new value—and revenue—adding functions while digitisation and optimising others.

I also believe that leadership should put themselves more often into the shoes of staff in addition to customers to be able to best understand the sentiment and act accordingly.

4.4 Building

The aviation industry has typically been extremely process oriented and risk averse, with big governance structures, also when it comes to running projects. A focus on results and agile behaviour is another thing which aviation can learn from tech companies in terms of how to run projects. Coupled with supporting new ideas and taking bold risks as part of the eco system, but abandoning when realising that it will not work out as predicted is crucial.

Methodologies such as design thinking [28] and Kanban help to design and run projects and achieve fast solutions in agile environments versus discussions without decisions over long periods of time and cumbersome governance structures for processes taking away empowerment of people. They can also help to ensure to draft processes which take account of future needs and allow flexibility rather than just reproducing similar approaches to today with new technology.

It is important that this approach is being understood and clearly becoming alive. It also involves taking some risks and creating a culture of trial and failure.

There are solutions in the market now which help to overcome some of the shortcomings of the old technology. If those solutions are adopted in a modular way, then gradually the unnecessary elements of the old systems can be phased out, and ultimately a truly state of the art proposition be in place, with well managed risk.

At the moment those innovative solutions are often ignored by airline people because they cannot yet imagine this new world. It is crucial for leadership to ensure that they take a leading role in guiding the organisation to do things differently. The biggest risk in the current environment is to not move.

4.5 Limitations and further considerations

Aviation is at a turning point. Changing consumer behaviour and customer expectations, rise of middle classes in developing economies, the global political landscape, environmental concerns and technological development lead to a dynamic environment and challenges never seen before.

Digitisation leads to a large scale of transformation across multiple aspects of business. It creates enormous opportunities, but also represents risks if not managed properly. The strategic implications for organisation, industry ecosystems and society have not yet been fully grasped by business leaders nor governments. Digitisation creates new challenges not yet fully understood. They include the pace of change never seen before, cultural change, the impact on society and identification of skills needed, outdated regulations, how to overcome legacy systems, the need for funding of both digital and physical infrastructure. Industry and Governments leaders need to take up the challenges in order to ensure that the potential value for society and industry can be leveraged. The question of the value of digitisation for aviation, travel and tourism is estimated to reach up to $305bn between 2016 and 2025 through increased profitability because of higher productivity, increased demand for products and services due to personalisation, sharing models and further improved perception of security. $100 billion (bn) of value are expected to migrate from traditional to new players in the industry (for example from traditional travel booking intermediaries to OTAs). $700 bn are the expected value for customers and wider society because of reduced environmental footprint, cost and time savings for travellers and safety and security improvements [29].

4.5.1 Customer experience

Travellers will expect a seamless experience tailored to their habits and preferences. Companies in the travel eco ecosystem along the customer journey will exchange data via secure technologies and continuously create insights. Travel will become frictionless and gradually blend with other daily activities. Digital technologies will augment the customer experience and the aviation workforce. Artificial intelligence (AI) and Machine Learning (ML) will help to turn data into insights and improve the customer experience, in the form of personalisation and chatbots, as well as take over specialist tasks of staff and transform the workforce. In addition, digital platforms, connected devices (Internet of things IoT), Virtual and Augmented Reality (VR/AR) and other technologies will allow for innovation, better customer experiences and increased efficiencies, and lead to a complete revision or erasion of legacy old processes. With digitisation of identify increased collaborative efforts need to be taken to ensure cyber security. The example of British Airways hacker attack on customer data in August 2018 is a good reminder of how real this threat is. Closely linked are fake news and fake revisions and evaluations of services via social media platforms.

4.5.2 Jobs and skills

The greatest societal impact of digitisation is probably the impact on the workforce and estimated to represent 1 in 11 jobs in the aviation and travel industry world wide according to the World Economic Forum study referenced above [31], potentially a number of 780,000 traditional job losses in the aviation and travel industry. Digitisation and new technologies will also mean displacement of current jobs in the industry, expected to be partially offset by next generation skilled jobs inside and outside aviation at the high and low end of the economy (for example in the area of robotics, Internet of Things (IoT), data analytics). All of these pose questions about future workforce which need to be addressed by industry and governments alike. New thinking is needed with regard to views on employment by society, concepts for next generation jobs and next generation occupation and pass time of people. Middle-level jobs that require routine manual and cognitive skills are the ones most at risk in terms of labour displacement and productivity effects [30]. Big legacy companies in particular struggle with the challenges of identifying new functions and redesigning organisation to integrate new and current functions in a way which suits the current dynamic environment [31]. Most departments have been run in silos, and staff fear about losing recognition and their jobs. Training programmes working with new technologies and helping to update relevant skills are required.

Top executives and board members have often been far away from digital and technological developments, and these areas have been specific entities in the organisation. It is a big challenge for these leaders to open up and learn fast about the relevant technologies they need to consider, what their set up should look like and strategic options and tactics how to get away from their legacy systems and processes. I have heard from many personal discussions with people within these organisations that many change activities do not go ahead as they should as leaders lack the insight and thus courage to decide to go ahead with radical changes when they get opposition from some people within the organisation.

4.5.3 Legacy systems

Airlines in particular but also other aviation and travel stakeholders face limitations in their activities and speed of transformation as they need to keep legacy systems running while developing new technology. They are afraid of the risk of changing the underlying legacy technology. Yet there are new technologies available now which could help to develop an environment for the "new world" for specific routes only as a test case and to get confidence while keeping the legacy systems running. Such a multi-speed approach to information technology (IT) requires strong leadership to move ahead successfully. Other limitations often encountered are the fact that technology and the knowledge going along with it had been outsourced by main players for many years. It is essential to develop some in-house knowledge and skills even to be able to understand and manage IT suppliers better. Innovation in terms of technology often happens much more with smaller suppliers in the aviation and travel world, which leads to the question of small versus large suppliers in the eco system. Aviation stakeholders have often feared being exposed to smaller suppliers, and bigger one-stop suppliers have fostered this fear, yet the current environment asks for new approaches and a critical review of the choice of a supplier in terms of innovation potential.

4.5.4 Regulation and legislation

The regulatory framework has a significant influence on transformation and can encourage or discourage the introduction of new technology. Innovation moves much faster than regulations and policy making, which means that Governments are forced to introduce regulations for nascent technologies. Concerted actions by industry leaders, regulators and policy makers are needed in order to maximise the value of digitisation in aviation, travel and tourism. The problem with fake news on social media reflects the risk of not embracing the new digital trends and not addressing the related opportunities and challenges. A series of actions for all participants in the ecosystem can be identified. They include the following according to the study by the World Economic Forum cited above [29]:

- Empower educational institutions to design curricula that help to prepare the next generation for the digital economy.

- Support the transition of the workforce with reselling current employees through training.

- A framework of rules for the operation of machines and AI systems is needed. Yet frameworks should remain flexible enough to not kill the innovative spirit but help to foster the development with guidelines and pro-active measures to address liability, safety, security and privacy of these new technologies.

- Transforming legacy systems into agile platforms with interoperability, enabling plug-and-play interactions between the partners in the ecosystem.

- Define a regulatory framework that defines the appropriate use of data, involving private, public and civil-society organisations.

4.5.5 Global political trends and economic evolution

International departures have more than doubled between 1996 and 2016, from 650 million to 1.45 billion, according to the world bank [32]. It appears that growth will continue. According to the World Economic Forum report on digital transformation for aviation, travel and tourism [29] global emerging markets will account for 70% of forecast share of global airline travel by 2034. Demographic developments play a key role in terms of growth and how fast new technology will be adopted. Regions in Asia, Africa and Latin America will drive a main part of this growth due to a rising middle class. Technology adoption may be speedier in developed countries though. Businesses will also face the challenge to manage experiences for travellers who are less used to technology.

Growth means that the aviation stakeholders need to adapt faster. But it also creates other problems in terms of overtourism and sustainability. This is further increased by additional cruise tourism. A number of places have started to tackle too many visitors. The authorities of the Philippines and Thailand have introduced a forced break for Boracay Island (Philippines) and Maya Bay (Thailand). Cinque Terre in Italy try an app with which tourists can see the number of people on the routes in real time. Machu Picchu in Peru turns to time slots. Jeju Island in South Korea faced almost 180 daily flights in 2017 and 15 million visitors, yet relief came not through the authorities but due to a Chinese ban not related to the underlying problem. Colombia's Caño Cristales site faces the challenge of balancing a delicate ecosystem with an unprecedented number of visitors. In a quite exceptional approach for a developing country they tackled this fast and introduced a set number of rules: no plastic bottles, no sunscreen or insect repellent in the water, no swimming in certain areas, no cigarettes, no feeding the fish. On arrival, visitors attend a briefing to make this completely clear. They are also training local tour guides and hosts [33].

Political tendencies to protectionism rather than continued globalisation as well as rising fuel prices could potentially have an impact on the growth forecast [34].

Other key considerations about the future evolution include

- How can stakeholders in the aviation and travel eco system ensure data security and comply with new data protection laws while incentivising customers to share personal data in exchange for tangible benefits, such as a hyper-personalised travel experiences. To what degree can personal data be securely and ethically used, and made interoperable across public and private stakeholders, to boost safety and security?

- The world of the hyper connected consumer is moving from physical to digital assets. Examples such as Uber, Amazon, Google, Apple, Expedia, Tesla, WhatsApp and more illustrate that the enterprise value of the future is about how well an organisation develops their digital assets for the benefits of customers and employees [35]. Is there a model for aviation to foster global collaboration and facilitating the sharing of company assets, to unleash the full potential of digital transformation, while also preserving the individual company's relevance in the battle for consumer mindshare? How will this impact on future investments in both physical infrastructure and digital technologies.

- How will the operating models of travel organisations change in a smart and connected world where the lines between online and offline are blurring, and physical assets turn to digital ones? How will this change the behaviour and expectations of individuals?

- Will it need completely new players in the market to finally push aviation and travel stakeholders towards more radical change? Similar as the low cost model gradually forced airlines and airports to change? Google now operates a large number of its own services, all branded accordingly, including Google Flights, Google Destinations and Google Hotels. Such improvements are already proving fruitful as more travellers turn to the Mountain View, California-based search company. According to the annual Portrait of American Travellers study from MMGY last year, 40% of travellers cite Google as their first source in booking trips. That's up 8 percentage points from the 2016 study [36].

5. Key conclusions

Aviation, particularly airlines' small profit margins and poor market capitalisation versus technology companies and other industries and increasing customer expectations are clear indicators that substantial change is needed to get fit for the twenty-first century. Airlines and airports have started the change process slowly, but a lot of digital transformation activities are ongoing in the meantime.

Main focus of activities is on customer experience improvements, cost efficiencies, better analytics and revenue optimisation as well as operational excellence. Internal and external innovation labs have been created to support the process, with more or less success so far. The most advanced companies have in-sourced or created at least some key parts of the software development activities.

Yet more drastic changes are still the exception, most of the activities are focused on creating workarounds based on decade old processes and systems. A lot of industry players either find it difficult to navigate in these stormy waters, or they prefer to stay ashore in the waters they know well and avoid any marks which indicate new ways because they cannot imagine that they will work.

It is critical for all board members and the whole leadership teams to have a deep understanding of the digital agenda, to ask the right questions and to drive the vision and the strategy. A big vision what the destination is and behaviour as prerequisites for branding and selling the trip to get the whole team work towards getting through stormy waters and test new ways to build the new world, even starting to build and show fast results are the main areas that still need to be completely fulfilled in many cases.

There are a lot of innovative start ups in the market, lots of opportunities to start drastic change. Disruptions and faster change will mean that the storm will get even stronger. Political changes and regulations, particularly with the increasing protectionist agenda of some countries are a risk for the foreseeable future in terms of expected growth.

Cost pressures above all due to increased labour and fuel cost but also in the area of aircraft cost are other main risks to be aware of. The latter could become bigger given the deals by Airbus with Bombardier and by Boeing with Embraer, which will restore the duopoly which the two giant manufacturers have had for many years. Both Bombardier with their C-series and Embraer with their E-series had started to compete directly with the smaller versions of Boeing and Airbus jets.

Technology will remain a key disruptor - but also a key enabler. If the big vision and behaviour start to get alive and are followed by branding and selling as well as building activities based on solution orientation, agile principles and the will to move forward and not remain in the past, then digitisation and current technological opportunities can open doors to do things previously thought impossible, creating seamless customer and staff experiences and creating endless new revenue and cost saving opportunities at the same time. Digitisation offers opportunities never seen before to shape the future. But industry leaders need to take up this chance and introduce the radical changes needed to create the potential value. Only the players who do this best will have a chance to survive and to compete successfully in the light of these dynamic technological changes and ever increasing customer demands. And competition is likely to increase strong players coming from originally other eco-systems such as Google, Amazon, Alibaba or others not seen before which will continue to move forward in the aviation and travel sphere.

Acknowledgements

I would like to thank IATA for having invited me as a jury member for their last hackathon in Kochi. Their hackathons contribute significantly to a change of mindset in the industry.

Thank you to the XXL Solutions team for research and empirical insight and to Hamburg Airport as the main sponsor for the think future event, which we have developed into the reference for innovation and transformation in the aviation and travel industry.

Conflict of interest

There is no conflict of interest to declare. Our strength is being an independent consultancy, which is very active in the digital transformation, innovation and start up travel and aviation arena.

Author details

Ursula Silling
XXL Solutions - Do Things Differently, Geneva, Switzerland

*Address all correspondence to: u@xxlsolutions.us

IntechOpen

References

[1] Elyatt H. CNBC. Ryanair Turns Customer-Friendly on Easy Jet Threat. 22/11/2013

[2] IATA Press Release No. 31: Corporate Communications (Sydney), Solid Profits Despite Rising Costs. 4/6/2018

[3] Forbes Global 2000 Publication. The World's Largest Public Companies, Edited by Halah Touryalai and Kristin Stoller with data by Andrea Murphy. 6/6/2018

[4] Business Insider. Zhang B. Delta's CEO Explains Why Airline Computers Fail and How Tech Will Change Flying. 17/12/2017. Available from: https://www.businessinsider.com/delta-ceo-explains-how-tech-will-change-flying-2017-12?IR=T

[5] IATA, Atmosphere Research Group. The Future of Airline Distribution. 2016-2021. Available from: https://www.iata.org/whatwedo/airline-distribution/ndc/Documents/ndc-future-airline-distribution-report.pdf. Airline Distribution Fundamentals - Current Problems, Disruptors and Future Perspectives

[6] Video 1: Live Stream Think Future - Hamburg Aviation Conference, YouTube Channel. 8-9 February 2018. Available from: https://www.youtube.com/playlist?list=PL5IfEpU_v0VCu5534ZWouJqSHU43VYHPS

[7] BBC Article. British Airways Boss Apologises for 'Malicious' Data Breach. 7/9/2018. Available from: http://www.bbc.co.uk/news/uk-england-london-45440850

[8] Journal Article. International Airport Review: Airlines and Airports to Invest US$33 Billion in 2017. 5/9/2017

[9] Business Travel Article. Jarvis H. Virgin to Launch New Loyalty Scheme with Delta. 21/8/2018. Available from:

https://standbynordic.com/virgin-to-launch-new-loyalty-scheme-with-delta/

[10] Future Travel Experience. Delta Invests $50m in RFID Baggage Tracking Technology. May 2016. Available from: https://www.futuretravelexperience.com/2016/05/delta-invests-50m-rfid-baggage-tracking-technology/

[11] Irish Times Article. Ryanair Wants to be "Amazon of Air Travel" With New Booking Option. 9/6/2016

[12] Article in SMBP Social Media for Business Performance, KLM: Using Social Media to Leverage "Service, Brand and Commerce. 2/4/2017. Available from: https://smbp.uwaterloo.ca/2017/04/klm-using-social-media-to-leverage-service-brand-and-commerce/

[13] staralliance.com: Star Alliance Creates Digital Service Platform with Accenture. 8/2/2018. Available from: https://www.staralliance.com/en/news-article?newsArticleId=DSP&groupId=20184

[14] Future Travel Experience Article. Heathrow Partners with Grab to Offer App-Based F&B Pre-order Service. September 2017. Available from: https://www.futuretravelexperience.com/2017/09/heathrow-partners-with-grab-to-offer-app-based-fb-pre-order-service/

[15] European Commission Website. Personalised Airport Systems for Seamless Mobility and Experience. 2015-2018. Available from: https://ec.europa.eu/inea/en/horizon-2020/projects/h2020-transport/aviation/passme

[16] Website Transport for London. Unified API - Transport for London. Available from: https://tfl.gov.uk/info-for/open-data-users/unified-api

[17] Business Architecture and Consultancy, Blog, Deutsche Bahn as a Digital Role Model. 2016. Available from: http://www.digitalsocialstrategy. org/bac/2016/12/09/deutsche-bahn-as-a-digital-role-model/

[18] Ad Age Article. Pasquarelli A. Overbooked: Expedia and Priceline Battle the Digital Duopoly. 19/3/2018. Available from: http://adage.com/ article/cmo-strategy/expedia-priceline-battle-digital-duopoly-airbnb/312769/

[19] Tnooz Article. Lufthansa to Add Surcharge to Every Booking Made via the GDS. 2/6/2015. Available from: https://www.tnooz.com/article/ lufthansa-to-add-surcharge-to-every-booking-made-via-the-gds/

[20] Passenger Self-service Article. Easy Jet Launches Connecting Flights Platform. 13/9//2017. Available from: https://www.passengerselfservice. com/2017/09/easyjet-launches-connecting-flights-platform/

[21] Munich Airport Website. A Humanoid Robot with Artificial Intelligence. February 2018. Available from: https://www.munich-airport. com/hi-i-m-josie-pepper-3613413

[22] Board of Innovation Blog. Khayati Y. Jobs in Innovation: Our Field Guide. 23/9/2015. Available from: https://www.boardofinnovation. com/blog/2015/09/23/ jobs-in-innovation-our-field-guide/

[23] easyJet Website - media centre. easyJet Signs Deal with Founders Factory to Create from Scratch and Accelerate Start ups to Innovate the Travel Sector. 16/10/2016. Available from: https://mediacentre.easyjet. com/en/stories/11200-easyjet-signs-deal-with-founders-factory-to-create-from-scratch-and-accelerate-startups-to-innovate-the-travel-sector

[24] Made by Many Blog. Braddock K. Innovation Labs - Best Practice, 16/11/2016

[25] GeekWire. Wong K. How Amazon Could Succeed in Travel: Researchers Issue a Warning to the Industry. 11/7/2018. Available from: https://www. geekwire.com/2018/amazon-succeed-travel-researchers-issue-warning-industry/

[26] CNBC. Kim T. Amazon Could Disrupt Online Travel Industry Next, Morgan Stanley Says. 9/3/ 2018. Available from: https://www.cnbc. com/2018/03/09/amazon-could-disrupt-online-travel-industry-next-morgan-stanley-says.html

[27] Business Insider Article. Zhang B. 'There's a Storm Coming', Emirates Boss Warns Airlines of a Looming Seismic Shift in Technology. 8/2/18 4:06 pm

[28] Article "Thisislarry". Reimagining Flight with People at the Center: How Design Thinking Can Change Air Travel. 2017. Available from: https://flytranspose.com/ reimagining-flight-with-people-at-the-center-how-design-thinking-can-change-air-travel-c9d6e2bb0d7d

[29] White Paper: World Economic Forum in Collaboration with Accenture. Digital Transformation Initiative. Aviation, Travel and Tourism Industry. January 2017

[30] Neufeind M, O'Reilly J, Ranft F. Work in the Digital Age. Challenges of the Fourth Industrial Revolution, Policy Network 2018

[31] M&S should look at Amazon tie-up, Says Marcus East Available from: http://www.bbc.co.uk/news/ business-44551664

[32] The World Bank. International Tourism, Number of Departures. 1996-2016. Available from: https://data. worldbank.org/indicator/st.int.dprt

[33] BBC News. Baker V. Tourism Pressures: Five Places Tackling Too Many Visitors. 16/4/2018. Available from: https://www.bbc.com/news/world-43700833

[34] Annual Economic Report, World Travel and Tourism Council (WTTC). Travel and Tourism, Global Economic Impact and Issues. 2017. Available from: https://www.wttc.org/-/media/files/reports/economic-impact-research/2017-documents/global-economic-impact-and-issues-2017.pdf

[35] Keynote Presentation at Think Future 18. Ghosh B. Leveraging Innovation Through Insight Into Other Industries, Think Future. 2018. Available from: https://www.hamburgaviationconference.com/publications/

[36] MMGY Study. Blount A. Portrait of American Travellers. 28/6/2017. Available from: https://www.mmgyglobal.com/news/news-2017%E2%80%932018-portrait-of-american-travelers

www.ingramcontent.com/pod-product-compliance
Lightning Source LLC
Chambersburg PA
CBHW081239190326
41458CB00016B/5843